# Science and Service Learning

A volume in
*Research in Science Education*
Dennis W. Sunal, Cynthia Szymanski Sunal, and Emmett Wright, *Series Editors*

## RESEARCH IN SCIENCE EDUCATION

Dennis W. Sunal, Cynthia, Szymanski Sunal, and Emmett L. Wright
*Series Editors*

*Reform in Undergraduate Science Teaching for the 21st Century* (2003)
edited by Dennis W. Sunal, Emmett L. Wright, and Jeanelle Bland Day

*The Impact of State and National Standards on K–12 Science Teaching* (2006)
edited by Dennis W. Sunal and Emmett L. Wright

*The Impact of the Laboratory and Technology on Learning and Teaching Science K–16* (2008)
edited by Dennis W. Sunal, Emmett L. Wright, and Cheryl Sundberg

*Teaching Science with Hispanic ELLs in K–16 Classrooms* (2010)
edited by Dennis W. Sunal, Cynthia, Szymanski Sunal, and Emmett L. Wright

*The Role of Public Policy in K–12 Science Education* (2011)
edited by George E. DeBoer

*Research Based Undergraduate Science Teaching* (2014)
edited by Dennis W. Sunal, Cynthia, Szymanski Sunal, Emmett L. Wright, Cheryl Mason, and Dean Zollman

*Science and Service Learning* (2016)
edited by Jane L. Newman, Dennis W. Sunal, Cynthia, Szymanski Sunal

# Science and Service Learning

*edited by*

**Jane L. Newman**
*The University of Alabama*

**Dennis W. Sunal**
*The University of Alabama*

**Cynthia Szymanski Sunal**
*The University of Alabama*

IAP

INFORMATION AGE PUBLISHING, INC.
Charlotte, NC • www.infoagepub.com

**Library of Congress Cataloging-in-Publication Data**

A CIP record for this book is available from the Library of Congress
http://www.loc.gov

ISBN: 978-1-68123-736-7 (Paperback)
978-1-68123-737-4 (Hardcover)
978-1-68123-738-1 (ebook)



Printed and bound by CPI Group (UK) Ltd, Croydon, CR0 4YY

# CONTENTS

# PREFACE TO THE SERIES

Science education as a professional field has been changing rapidly over the past two decades. Scholars, administrators, practitioners, and students preparing to become teachers of science find it difficult to keep abreast of relevant and applicable knowledge concerning research, leadership, policy, curricula, teaching, and learning that improve science instruction and student science learning. The literature available reports a broad spectrum of diverse science education research, making the search for valid materials on a specific area time-consuming and tedious.

Science education professionals at all levels need to be able to access a comprehensive, timely, and valid source of knowledge about the emerging body of research, theory, policy and practice in their fields. This body of knowledge would inform researchers about emerging trends in research, research procedures, and technological assistance in key areas of science education. It would inform policy makers in need of information about specific areas in which they make key decisions. It would also help practitioners and students become aware of current research knowledge, policy, and best practice in their fields.

For these reasons, the goal of the book series, Research in Science Education, is to provide a comprehensive view of current and emerging knowledge, research strategies, and policy in specific professional fields of science education. This series presents currently unavailable, or difficult to gather, materials from a variety of viewpoints and sources in a usable and organized format.

*Science and Service-Learning*, pages vii–viii

Each volume in the series presents a juried, scholarly, and accessible review of research, theory, and/or policy in a specific field of science education, K–16. Topics covered in each volume are determined by current issues and trends, as well as generative themes related to up-to-date research findings and accepted theory. Published volumes will include empirical studies, policy analysis, literature reviews and positing of theoretical and conceptual bases.

# PREFACE

*Science and Service Learning*, RISE Volume 7, focuses on teaching science as a multifaceted research-based value added strategy designed to improve academic engagement, civic responsibility, and resiliency for students "at-risk" to drop out of school. In addition, science and service-learning is recognized as an effective strategy for motivating students to attend college, plan for STEM careers, and encourage students and adults to collaborate as partners to improve their communities.

Goals are selected to promote dialogue in the research community for sharing research on teaching conducted in science and service-learning in K–16 contexts. This strategy may include impact of science and service-learning on:

- Student science academic engagement and academic achievement
- Teacher instructional strategies
- Science curricula adaptation or development
- Social/civic responsibility and collaboration between students and community partners
- Resiliency of "at-risk" students
- STEM career interest
- Workforce readiness

Our nation faces a dropout crisis. One-third of our total students and one-half of our African American and Hispanic students fail to finish high school in four years. Of those who do finish, a significant number of U.S. students,

*Science and Service-Learning*, pages ix–x

especially those living in high poverty areas, lag behind their international peers in math and science. Experts in business and education argue that unless America renews a commitment to excellence and development of scientific talent, the long-term prosperity of our nation may be at risk.

A majority of school dropouts have reported that had they been provided engaging instruction in high school with real-world learning to make school more relevant (experiences like service-learning), they would have been more likely to have finished school. Having been recognized as an effective strategy for inspiring student learning, active learning, and engagement in school, service-learning has been credited as an instructional strategy for improving high school graduation rates, motivating students to finish college and engage in careers about which they are passionate. At the same time, service-learning has benefitted communities as students have worked side-by-side with adult partners including community leaders, researchers, parents, teachers, and other students.

In the chapters that follow, a number of researchers, teaching and curriculum reformers, and reform implementers discuss a wide range of issues. The volume should be considered as a first step in thinking through what science and service learning might address in teacher preparation programs, in-service professional development programs, and in classrooms throughout the United States. To facilitate our thinking about research-based science and service learning each chapter in the volume is organized around five common elements:

1. A significant review of research in the issue or problem area.
2. Themes addressed are relevant for the teaching and learning of K–16 science and service-learning.
3. Discussion of original research by the author(s) addressing the major theme of the chapter.
4. Bridge gaps between theory and practice and/or research and practice.
5. Concerns and needs are addressed of school/community context stakeholders including students, teachers, parents, administrators, and community members.

**—Jane L. Newman**
**Dennis W. Sunal**
**Cynthia S. Sunal**
The University of Alabama

CHAPTER 1

# IMPACT OF SCIENCE AND SERVICE LEARNING ON MIDDLE SCHOOL STUDENTS' RESILIENCE

**Jane L. Newman**
***University of Alabama***

**Junfei Lu**
***University of Alabama***

## ABSTRACT

This study examined the influence of an innovative science and service-learning instructional program (Science in Action [SIA]) on students' measure of resilience. The program aimed to transform traditional science classrooms into engaging learning environments. Some 20 middle schools serving students from high-poverty families participated in SIA the program. Over 100 science teachers and 120 community partners worked to incorporate science, technology, engineering, and math (STEM) problem-based learning pedagogy into science instruction to teach students to solve real problems in schools and communities. Two analyses were conducted. The first analysis examined students' pre- and post-SIA implementation test scores on resilience from

*Science and Service Learning*, pages 1–18

four middle schools. The second analysis targeted the impact of the quality of school SIA implementation on students' resilience after they completed the programs.

*If I hadn't been involved in this science service-learning program during middle school, I would have dropped out of school a long time ago. That outstanding experience literally saved my life, and I wouldn't be graduating with my senior class, going to college and planning a future in technology!*

—Juan

When Juan made the statement above, he was a graduating senior in a high school where subgroups of students, including economically disadvantaged and others, were likely not to finish high school. Such schools produce one third of the numbers of young people who drop out before finishing high school within 4 years (GradNation, 2016) Juan was an exception. He had experienced a quality science-based service-learning project in his middle school science classes that changed his life and his future goals.

## SERVICE LEARNING AND SCIENCE IN ACTION INITIATIVE

Service learning is a method of teaching and learning that offers a unique opportunity for young people to use what they learn in the classroom to solve real-life community problems and to develop real-world services benefiting people. This method integrates academic study with meaningful service and reflective practice, strengthens schools and communities, and can build academic and civic engagement and resiliency to adversity (YSA, 2013). Learning situations can be designed to teach students to become producers of knowledge and services, not just consumers of information (good test takers).

One initiative experiencing some success was housed at a large southeastern research university. The Science in Action (SIA) initiative was a project combining science and service learning as an effective pedagogical technique. The National Research Council (NRC) posited "inquiry into authentic questions generated from student experiences is the central strategy of teaching science... focus[ing] inquiry predominantly on real phenomena in classrooms, outdoors, or in laboratory settings" (as cited in Jordan, 2005, p. 43). Inquiry-based service learning can provide youth the opportunity to engage in problem-based, hands-on, collaborative, higher-order thinking skills and processes with their peers and community partners. Students practice important thinking skills and processes within an academic context and realize the purpose for their learning because they make an impact that improves problems in their schools and communities.

Students learn to recognize and solve tough problems, which can become passions leading to future careers in science and community engagement.

The SIA program selected, funded, and led 20 subgrantee Alabama and Georgia middle schools (grades 6–8) in engaging students in problem-based, service-learning projects. The innovative program funded by the Corporation for National and Community Service (CNCS) included more than 6,000 middle school at-risk students (eligible for at least 50% free or reduced-cost lunch), and 106 science teachers with assistance from 120 community partners. Science teachers integrated problem-based service-learning pedagogy into science instruction to help students solve real problems in schools and communities.

This project was unique because it included "cognitive apprenticeship" (Collins, Brown, & Newman, 1989). SIA provided professional development for participating middle school science teachers to assist students in improving academic engagement, with the aim of building resilience to adversity. Teachers worked with colleagues and experts developing their pedagogical knowledge base and ability to facilitate science-related service-learning projects addressing real problems in their schools and communities. Teachers worked with each other, building collegial support within each school, across schools, and across the two states. The focused theme of science-related issues enabled teachers to communicate to students the role of their discipline within the context of 21st century life.

## Professional Development

The initial professional development included 4 days of training for teachers. Topics of the training were problem-based learning, service learning, the Talents Unlimited Thinking Skills Model (Schlichter, 1987), professional learning communities, and technology. One of the approaches in the professional development was cognitive apprenticeship involving a group of learners exchanging roles from teacher to learner and back again. This approach focuses on intervention and collaboration and has been shown to be an effective means of creating change (Collins et al., 1989). The apprenticeship was designed to change implicit everyday knowledge to explicit informed practical knowledge through shared reflection and action research. Teachers were closely monitored throughout the life of the program. Decisions and program improvements were made based on ongoing data analysis asking the questions, Are students improving in creative and critical thinking skills and problem-based learning? Are teachers benefiting from professional learning communities? Are students becoming more engaged in their science classes? Are students making connections between science-related problems in their schools and communities?

Are students developing civic responsibility? Are students becoming more resilient to adversity?

## Citizenship

The "Civic Mission of Schools" (Levine, 2003) characterizes responsible and competent citizens as informed and thoughtful participants in their communities through addressing social, political, cultural, and religious interests; accomplishing public goals; and working for rights and welfare of others. Service learning, as compared to volunteerism, involves research inquiry as well as investigation of and solutions to real-world problems, providing students with opportunities to learn important lessons in civic responsibility. The more opportunities youth have to engage in this type of learning, the more likely they will continue to take action as responsible adults (Renzulli & Reis, 2014).

## Disadvantaged Youth

This project focused on disadvantaged youth's transformation from having been served to serving others. Too often young people are seen as the recipients of service (Newman & Dantzler, 2015). Instead, they can learn to see themselves—and can be seen—as leaders and assets who can make powerful contributions to their schools and communities (Newman, 2005; Renzulli & Reis, 2014). Strategies for engaging youngsters include

1. View all youth as resources, not liabilities;
2. Diversify leadership to give all young people a chance to learn;
3. Create a group environment where all feel safe and wanted;
4. Invite all to have input on addressing issues/concerns;
5. Increase responsibilities to realize successes; and,
6. Have caring adults to mentor students (Roehlkepartain, 2007).

## Diversity

Service-learning projects can bring together individuals from diverse backgrounds. Subgrantee local education agencies (LEAs) were required to document that at least 50% of their students received free or reduced lunch. Each subgrantee selected was required to demonstrate how the respective service-learning project would connect participants of different backgrounds including race, gender, religion, socioeconomic status, and age.

## Community Partners

Developing community partnerships was a critical strategy in implementing a comprehensive, high-quality service-learning initiative because it enabled us to serve larger populations, make better use of resources, avoid duplication of effort, learn from one another, and solve problems more effectively. Community partners worked with youth to develop reciprocal, sustainable relationships assuming a vested interest in assisting subgrantees to successfully pursue their service-learning endeavors. State departments of education in both Georgia and Alabama were integral parts of our network of partners. The Vice Chancellor of Academic Affairs at the University of Alabama (UA) and the Director of the Alabama Mathematics, Science and Technology Education Coalition (AMSTEC) served on the SIA Advisory Board to assist in networking with business partners.

## Sustainability

A key consideration in designing the project was sustainability. We chose to provide research-based staff development for existing personnel who work with youngsters. Initial training at retreats, incubation of service-learning concepts and constructs, and interaction and developing strong reciprocal relationships and partnerships with multiple local and state organizations contributed to sustainability. Teachers learned to integrate service learning into their science units, as well as institutionalized practices and principles in their programs. High-quality technical assistance was provided to ensure the service-learning practices were continued. Additional funding and community involvement were addressed in strategic plans to develop policy to ensure that programs continued after the funding ended.

Sustainability was attained through the support system for this project already fully in place in Alabama. The Alabama Mathematics, Science and Technology Initiative (AMSTI) had been operational in Alabama for several years and is fully funded by state dollars. Teachers included in this program had received at least 2 years of related AMSTI training and continued to be trained in collaboration and coordination of funding sources (federal, state, and local) to maximize potential benefits for students. Support from the Alabama State Department of Education (ALSDE) included technical assistance from the Learn and Serve America (LSA) state coordinator as well as technical assistance from the AMSTI staff. The LSA coordinator assisted with training for the subgrantees on how to integrate service learning into the classroom and provided technical assistance for participants on using existing federal funds to ensure sustainability.

Service-learning projects have the potential to engage schools and youth service organizations into long-term partnerships. The service-learning projects provided access to hands-on, real-world science learning and civic stewardship for youngsters through school and community projects. Other sustainability factors included integration with curriculum, standards, and organizational missions; increasing organizational capacity through professional learning communities and continued technical support; continuing to clarify roles and responsibilities among partners; and assisting with installing policy to sustain the projects after funding had ended.

## SERVICE-LEARNING STANDARDS FOR QUALITY PRACTICE

Students who participate in service-learning experiences can display significant increases in academic, civic, and resilience outcomes (Billig, 2002; Furco, 2002; Furco & Root, 2010). In 2008 the National Youth Leadership Council (NYLC, 2008) published a set of eight standards, K–12 Service-Learning Standards for Quality Practice (see Table 1.1), to improve the uniformity and increase the rigor of the service-learning strategy as an instructional practice. There is little research literature related to service-learning

**TABLE 1.1 National Youth Leadership Council K–12 Service Learning Standards for Quality Practice**

| K–12 Service Learning Standards for Quality Practice | |
|---|---|
| Duration and Intensity | Service learning has sufficient duration and intensity to address community needs and meet specified outcomes. |
| Meaningful Service | Service learning actively engages participants in meaningful and personally relevant service activities. |
| Link to Curriculum | Service learning is intentionally used as an instructional strategy to meet learning goals and/or content standards. |
| Progress Monitoring | Service learning engages participants in an ongoing process to assess the quality of implementation and progress toward meeting specified goals and uses results for improvement and sustainability. |
| Reflection | Service learning incorporates multiple challenging reflection activities that are ongoing and that prompt deep thinking and analysis about oneself and one's relationship to society. |
| Youth Voice | Service learning provides youth with a strong voice in planning, implementing, and evaluating service learning experiences with guidance from adults. |
| Partnerships | Service learning partnerships are collaborative, mutually beneficial, and address community needs. |
| Diversity | Service learning promotes understanding of diversity and mutual respect among all participants. |

programs that includes high-quality implementation of service-learning standards. Researchers, however, have indicated that increased levels of knowledge and understanding in academic engagement and school resilience are more likely to be reached when experiences meet high-quality standards (Billig, 2010; Billig, Root, & Jesse, 2005; Furco & Root, 2010; Melchior & Bailis, 2002.

Quality service learning first was defined as demonstrating each of the following eight research-based quality indicators: duration and intensity, link to curriculum, partnerships, meaningful service, youth voice, diversity, reflection, and progress monitoring (NYLC, 2008). These standards provide a gauge for educators and researchers to measure components of quality service learning in more quantifiable terms and are most critical for producing positive student outcomes. Teachers and students were taught to integrate the eight standards. The following section details the application of the standards with the SIA subgrantees.

## Duration and Intensity

Middle school students participated in a service-learning project for at least one year over two semesters, for approximately 70–80 hours. Teachers participated in the project's program for 2 to 3 years, including attending ongoing professional development sessions, meeting in local and online professional learning communities, and planning with community partners.

## Link to Curriculum

Projects and activities were tightly aligned with state and national course of study objectives. Instruction integrated technology, mathematics, and higher-order thinking skills through a problem-based, inquiry-based approach (Renzulli & Reis, 2014) to science. Through use of the Talents Unlimited Model (Newman, 2005), students applied critical and creative thinking skills to plan and implement projects, including productive thinking, decision making, communicating, forecasting, planning, applying, analyzing, synthesizing, and evaluating.

## Meaningful Service

Middle school students selected real-world science problems to investigate and solve through planning and creating a project to provide a service to their school or community. Teachers guided by identifying overarching topics. Students contributed input and took leadership at each stage of project development (i.e., selection of project subtopics and types development of Student Action Plans, etc.) A sample of projects implemented in sixth through eighth grades included creating greenhouses and outdoor classrooms to produce healthy produce to feed the local community, working with community experts to conduct water testing to determine if water quality might be linked to a county's high cancer rate, planning a health awareness community fair to encourage adoption of healthier lifestyles and prevention of diabetes or obesity, working with special-needs students to participate in service-learning projects, and working with alternative school students whose curriculum included a volunteer landscaping component.

### Progress Monitoring

In addition to teacher ongoing assessment, students regularly monitored their progress with project quality and meeting timelines. Student action plans charted progress. Students and teachers used Liptrot's (2010) *K–12 Service-Learning Standards for Quality Practice Evaluation Tool* to evaluate and make modifications to their action plans. Regular conferences between students and teachers were held to ensure projects were meeting quality standards and projected timelines.

### Reflection

Students applied reflection strategies during every stage of the process through participation in technology-infused, higher-order thinking activities. Through implementation of the Talents Unlimited Model (Newman, 2005), students used the following skills to make connections while simultaneously attaining academic skills: productive thinking, decision making, planning, communicating, and forecasting. Students also applied numerous Web 2.0 tools (i.e., blogs, wikis, videos, mind maps, and digital storytelling) to document and to communicate their personal and collective reflections. Projects were posted to a central website enabling teachers and students to collaborate about new projects and learn from one another.

### Partnerships

The University of Alabama's College of Education served as a partner to all participating schools, providing monetary resources, professional development, and progress monitoring. Each school worked with an average of seven community partners, including local businesses and nonprofit organizations, who supported teachers and students in project development through both time and monetary resources, and, who in turn, benefited from the results of the services created. Most of the 20 schools involved in SIA had already formed community partnerships before they were funded by the project.

### Youth Voice

Throughout their projects students played a leading role in planning, implementing, and evaluating. The students applied higher-order thinking skills and science problem solving strategies. They worked in an environment of trust and open ideas that fostered leadership, planning, decision making, and evaluating the quality and effectiveness of the service-learning experience (NYLC, 2008).

## RESILIENCY

Children who overcome adversity despite many obstacles are described as being resilient (Benard, 2004). Most researchers agree that resiliency can evolve in children who have innate contributing attributes. These innate attributes can be further developed to foster resiliency by the action of environmental factors, especially in school environments.

### Innate Attributes

The foundation from which to start building students' resiliency is for caring adults in schools to believe that every student has innate resilience. Contributing to innate resilience are attributes, including social competence, a sense of identity and autonomy, and a sense of purpose including goal direction and belief in a bright future (Benard. 1996). Educators can help students develop particular innate traits that enable them to transform from service receivers to service givers if the traits interact with positive environmental protective factors (Benard, 2007).

### Environmental Protective Factors

According to Rutter (1984), when dysfunctional families fail to provide children with meaningful relationships built on positive environments in which at-risk children can succeed, schools and communities are critical. In schools youth can experience mutually caring relationships that give meaning to their lives, for example, a reason for caring and commitment to serve others (McLaughlin, Irby, & Langman, 1994; Werner & Smith, 1992).

Although resilience research in K–12 education is relatively new, it focuses on practices fostering healthy development that can prevent failure in school in spite of risk factors and adversity (Werner, 1984). Most resilience researchers agree at least three environmental factors are critical in the development of school resilience and dropout prevention (Benard, 2014):

- Close, caring relationships, including trust;
- High expectations, challenge, positive expectations; and
- Contributing one's talents to the community, especially in middle school and adolescence.

By becoming more aware of schools' environmental protective factors and making a commitment to strengthen them, teachers and schools can play a significant role in developing resilience (Henderson, 2013). Schools do so by providing opportunities promoting real responsibility and real work (McLaughlin et al., 1994). Extracurricular programs, such as service-learning experiences, incorporate duration and intensity, not just short acts of kindness in the classroom. These long-term experiences provide opportunities nurturing self-esteem and giving students experiences of required helpfulness, an environmental factor that may help to engage students to develop school resilience (Bridgeland, DiIulio, & Morison, 2006).

### Purpose of This Study

We examined the relationship between standards-based science, service learning, and resilience (SIA). The two research questions investigated were

1. Does students' measured resilience change significantly after students participate in a service-learning program?
2. Does the quality of implementation of service learning impact students' measured resilience?

## METHOD

The sample for the first analysis (to answer the first research question) included participants from four schools providing valid data on both pre- and posttest scores of students' resilience measure. There were 411 students composing the pretest dataset. Among them, 157 (38%) were male and 251 (61%) were female students, with three missing values; 199 (48%) in sixth grade, 80 (20%) in seventh grade, and 125 (30%) in eighth grade, with seven missing values. Regarding ethnicity, 210 (51%) were Caucasian, 120 (29%) black, 55 (13%) Hispanic, and eight (2%) others, with 18 missing values. The posttest dataset represented 430 students. Among them, 206 (48%) were male and 223 (51.9%) were female students, with one missing value; 93 (22%) in sixth grade, 209 (49%) in seventh grade, and 126 (29%) in eighth grade, with two missing values. Regarding ethnicity, 276 (64%) were Caucasian, 42 (10%) black, 83 (19%) Hispanic, and 15 (4%) others, with 14 missing values.

The sample for the second analysis (to answer the second research question) included 1,483 students based on the entire posttest measure of resilience. Among them, 725 (49%) were male and 747 (50%) were female students, with 11 missing values; 390 (26%) in sixth grade, 850 (57%) in seventh grade, and 230 (16%) in eighth grade, with 13 missing values. Regarding ethnicity, 873 (59%) were Caucasian, 310 (21%) BLack, 184 (12%) Hispanic, and 48 (3%) others, with 68 missing values.

The *Learn and Serve America: Resiliency Student Survey (Grades 6–12)* (Billig, 2008) was used in the study (see Table 1.2). This survey is a 17-item scale to measure students' level of resilience. The scale is composed of two

**TABLE 1.2 Learn and Serve America: Resiliency Student Survey (Grades 6–12)**

*Personal Resiliency was measured by six statements rated on a 4-point Likert-type scale from strongly disagree to strongly agree. The questions were:*

1. I enjoy working together with students my age.
2. I can work with someone who has different opinions than mine.
3. I stand up for myself without putting others down.
4. I plan to graduate from high school.
5. I plan to go to college or some other school after high school.
6. I have goals and plans for the future.

*School Resiliency was measured by eleven statements rated on a 4 point Likert-type scale from strongly disagree to strongly agree. The questions were:*

1. I do things at school that make a difference.
2. At least one teacher believes I will be a success.
3. The schoolwork I am assigned is meaningful and important.
4. I do interesting activities at school.
5. At school I help decide things like class activities or rules.
6. At least one teacher listens to me when I have something to say.
7. At least one teacher or adult at school notices when I am not there.
8. At least one teacher or adult at school always wants me to do my best.
9. At least one teacher at school listens to students' ideas about how to improve the school.
10. At least one teacher or adult at school really cares about me.
11. The things I am learning in school will be important for my future.

theoretical subdomains: school-based (11 items) and students'-based (6 items) of resilience. The Cronbach alphas for both subdomains were .82 and .93 for school and student-based ones, respectively. This scale is still a new scale, and more studies are needed to further validate its factor structure. In other words, the two-factor model needs to be further investigated to ensure its validity. We therefore used the total score across 17 items to represent an individual's measured degree of resilience. The Cronbach alpha for 17 items in the current study was .92. Each item is answered on a 4-point Likert scale (from 1 = strongly disagree to 4 = strongly agree). The higher the score achieved, the greater resilience. Sample items were "At least one teacher or other adult at school believes I will be a success" for school-based resilience, and "I stand up for myself without putting others down" for student-based resilience.

Prior to the research, the study proposal was approved by the university institutional review board (IRB). Pre- and postcourse data were collected using an online SurveyMonkey data-collection tool. Students in the service-learning projects at each of the 20 schools were expected to complete an online survey at the beginning and at the end of the school year. The survey measured their academic engagement, civic engagement, and level of resilience. For this chapter only the data from the resilience survey was used. The SIA program administrators relied on the individual schools to have students complete the pre-and postcourse data in a timely manner. Some schools' teachers, however, asked students to enter the precourse data well after the service-learning courses had begun. In addition, 10 of the 20 schools did not collect precourse data. Due to the elimination of data not meeting the requirements, only four schools provided valid data on both pre- and postcourse measures, but almost all schools provided the postcourse measure on resilience. Since all identifiable information was prohibited in this study, there was no identifier to determine whether or not scores on pre- and postcourse measures were based on the same student. This limited our analysis of score change on the resilience measure before and after the SIA implementation.

## DATA ANALYSIS

The first analysis was based on schools providing test scores of participants (students) on the resilience measure before and after the implementation of the SIA program. Based on valid data from four schools, we examined descriptive statistics to formulate a sense of students' pre-and postcourse measure on resilience. Other than descriptive analysis, we also checked inferential statistics (e.g., *t*-test, ANOVA) within pre- and posttests to examine the difference regarding the resilience measure among students based on their gender, grade, and ethnicity (Caucasian versus non-Caucasian). Using these analyses we endeavored to examine the impact of the SIA on students' resilience.

In the second analysis, we focused on posttest scores of resilience. Particularly, we paid attention to the impact of the quality of SIA program implementation on the resilience measure. Two evaluators (the first author and an external evaluator) independently rated the quality of each school's service-learning program, using Liptrot's (2010) *K–12 Service-Learning Standards for Quality Practice Evaluation Tool*, which includes the eight Quality Service-learning standards and indicators. Then both evaluators met together to discuss their results. They agreed on 85% of the ratings. For the remaining 15%, they exchanged viewpoints and reached consensus. In the end, schools were separated into high-quality versus low-quality SIA program implementation. We then compared students' resilience scores accordingly.

## RESULTS

For the first analysis, the outcomes showed participants' average score on the resilience measure was 55.76 ($SD = 6.90$, $n = 411$) on the pretest, and 55.56 ($SD = 8.18$, $n = 430$) on the posttest. The pretest indicated students' resilience scores had no significant difference regarding gender, $t(406) = -1.57$, $p = .12$; and ethnicity, $t(391) = .54$, $p = .58$. However, the ANOVA test showed that there was a significant difference regarding resilience among students in sixth, seventh, and eighth grades, $F(2, 401) = 6.96$, $p = .001$, $\eta^2 = .03$. In the follow-up analysis, we employed the Scheffé test. The strengths of this test include it is a robust test given unequal sample size, and that this test is comparatively "insensitive to departures from normality and homogeneity of variance" (Hays, 1994, p. 455). The result showed a decreasing trend among students' resilience scores, with sixth grade the highest and eighth grade the lowest. The difference between students in sixth and eighth grades was statistically different.

The students' posttest resilience scores had no significant difference regarding race, $t(414) = .92$, $p = .36$; and grade, $F(2.\ 425) = 2.91$, $p = .06$. There was a statistically significant difference, nonetheless, between male and female students: $t(427) = -3.5$, $p = .001$, $d = .3$. Table 1.3 indicates, means (M) and standard deviations (SD) of participants' resilience measure in pretest and posttest across gender, ethnicity, and grade.

For the second analysis, the mean score on resilience of participants from low implementation schools was 53.4 ($SD = 8.86$, $n = 478$). In comparison, the mean score on resilience of participants from low implementation schools was 54.8 ($SD = 8.33$, $n = 1005$). The independent *t*-test had a $t(1481) = -2.84$, $p = .005$, $d = .16$.

**TABLE 1.3 Students' Resilience Measure Before (Precourse) and After (Postcourse) the Implementation of SIA Programs—Sample Size (*N*), Mean (*M*), and Standard Deviation (*SD*)**

| | Resilience | | |
|---|---|---|---|
| | *N* | *M* | *SD* |
| **Precourse** | | | |
| *Gender* | | | |
| Male | 157 | 55.03 | 6.82 |
| Female | 251 | 56.13 | 6.92 |
| *Grade* | | | |
| 6th | 199 | 56.88 | 6.03 |
| 7th | 80 | 55.48 | 7.54 |
| 8th | 125 | 54.30 | 7.41 |
| *Ethnicity* | | | |
| Caucasian | 210 | 55.88 | 7.03 |
| Non-Caucasian | 183 | 54.49 | 6.90 |
| **Postcourse** | | | |
| *Gender* | | | |
| Male | 206 | 54.15 | 9.08 |
| Female | 223 | 56.88 | 7.03 |
| *Grade* | | | |
| 6th | 93 | 56.58 | 7.64 |
| 7th | 209 | 55.99 | 7.92 |
| 8th | 126 | 54.15 | 8.86 |
| *Ethnicity* | | | |
| Caucasian | 276 | 55.86 | 7.91 |
| Non-Caucasian | 140 | 55.07 | 8.61 |

## DISCUSSION

The overarching research question of this study focused on the effectiveness of the SIA program with regard to influencing students' resilience measure. Based on the findings retrieved from pre- and postcourse tests, there was no obvious change regarding students' scores on resilience before and after the implementation. In general, students on average maintained good resilience before and after participating in the SIA program. Such a level of resilience did not vary significantly in terms of gender, ethnicity, and grade. Although the above results show some statistically significant outcomes, the corresponding effect sizes were small. Therefore, we posit the effect of the SIA program to increase students' resilience was not observed in the current study. There are several factors that might have contributed to the current nonsignificant results. First, participants in their precourse test had demonstrated a good level of resilience, which might leave little space for them to further improve. This leads to our second concern. Resilience might be a trait that takes time to improve, especially from a good state to a better one. Even though the current study lasted for a year, it still might have not been long enough to capture the slow change process of students' resilience, if such occurred. The third issue concerns the scaling. Since the resilience scale only applies a 4-point Likert scale (strongly disagree, disagree, agree, and strongly agree), the scale might not be intricate enough to reflect students' resilience change.

Regarding the effect of the quality of SIA program implementation on students' resilience, the results showed a significant difference between students from low-quality and high-quality implementation schools. In particular, students from high-quality implementation schools demonstrated a slightly greater resilience than those from low-quality implementation schools. Due to the posttests' nature as well as the low effect size, we could not attribute this effect only to the quality of implementation or consider this effect as having significantly practical implication. This result, nonetheless, raises our attention to the quality of implementation, which might have a direct impact on students' resilience and may be worthy of further study.

## IMPLICATIONS

While the results of this study cannot be construed to argue that high-quality service-learning programs led to students' increased level of resilience, the results revealed a possible connection between quality service learning and good resilience maintenance. The data collection process from the project's evaluation was a major barrier impeding us from conducting a rigorous statistical analysis of the effect of SIA programs on students' resilience change. Some schools did not collect precourse data and others collected such data in a delayed manner, rendering those data invalid. Although the program may have the potential to enhance students' resilience, the undesirable execution of the data collection directly impacted the robustness of the analysis. We note that during the professional development stage prior to the implementation of the program, the primary focus was on introducing teachers to the SIA program and its implications for student learning, while paying little attention to helping teachers develop the basic research capacity critical to data collection.

More research is needed to identify components critical in service-learning design to promote students' resiliency. Long-term extracurricular programs (service learning) that uphold the *K–12 Service-Learning Standards and Indicators of High Quality Service Learning* (NYLC, 2008) may help students develop resilience. Teachers and administrators can identify community projects providing a fit for students based on their interests. These experiences may help to shape self-esteem, contribute to self-efficacy and transformation from at-risk to resiliency. Teachers, however, must be intentional in incorporating the standards into curriculum. In addition, the standards must be constantly monitored to determine if they provide quality instruction and reflection. As new rigorous programs are conducted, educators may recognize the importance of implementing high-quality programs and their effects on student resilience in addition to academic engagement and civic engagement. Therefore, quality service-learning initiatives like Science in Action may make significant differences in the lives, schools, and communities of students like Juan.

## AUTHOR NOTE

Jane L. Newman, College of Education, University of Alabama at Tuscaloosa. This research was supported by a grant from the Corporation for National and Community Service. Correspondence concerning this chapter should be addressed to Jane Newman, College of Education, University of Alabama, Tuscaloosa, AL, 35487. Email: jnewman2@ua.edu.

## REFERENCES

Benard, B. (1996). *Risk to resiliency: What schools can do.* Retrieved from http://www.tanglewood.net/projects/teachertraining/Book_of_Readings/Benard.pdf

Benard, B. (2004). *Resiliency: What we have learned.* San Francisco, CA: WestEd.

Benard, B. (2007). Gender and resilience. In N. Henderson (Ed.), *Resilience in action: Practical ideas for overcoming risks and building strengths in youth, families, and communities* (pp. 22–38). Solvang, CA: Resiliency in Action.

Bernard, B. (2014). The foundations of the resiliency framework. Resiliency in action (2nd ed.). Santa Barbara, CA: Santa Barbara Web Design.

Billig, S. H. (2002). Support for K–12 service learning practice: A brief review of the research. *Educational Horizons, 80,* 184–190.

Billig, S. H. (2008). *Learn and serve America: Academic engagement student survey, civic engagement survey, resiliency student survey (grades 6–12).* Denver, CO: RMC.

Billig, S. (2010, Summer/Fall). Why service learning is such a good idea: Explanations from the research. *Colleagues, 5*(1), 9–11 Retrieved from https://

www.ceesa.org/phocadownload/servicelearning/part_4_supplements_articles/5_colleaguesmagazine_servicelearning.pdf

Billig, S. H., Root, S., & Jesse, D. (2005). The relationship between quality indicators of service learning and student outcomes: Testing professional wisdom. In S. Root, J. Callahan, & S. H. Billig (Eds.), *Advances in service learning research: Improving service learning practice: Research on models to enhance impacts* (pp. 97–115). Greenwich, CT: Information Age.

Bridgeland, J. M., DiIulio, J. J., Jr., & Morison, K. B. (2006). The silent epidemic: Perspectives of high school dropouts. *ERIC.* Retrieved from http://eric.ed.gov/?id=ED513444

Collins, A., Brown, J. S., & Newman, S. E. (1989). Cognitive apprenticeship: Teaching the craft of reading, writing and mathematics! In L. B. Resnick (Ed.), *Knowing, learning, and instruction: Essays in honor of Robert Glaser* (pp. 453–494). Hillsdale, NJ: Erlbaum.

Furco, A. (2002). Is service-learning really better than community service? A study of high school service. In A. Furco & S. H. Billig (Eds.), *Service learning: The essence of pedagogy* (Vol.1, pp. 23–50), Greenwich, CT: Information Age.

Furco, A., & Root, S. (2010). Research demonstrates the value of service learning. *Phi Delta Kappan, 91*(5), 16–20.

GradNation. (2016). *Building a grad nation report: Progress and challenges in raising high school graduation rates.* Retrieved from http://www.gradnation.org/report/2016-building-grad-nation-report?_ga=1.242620097.2118565261.1469477585

Hays, W. L. (1994). *Statistics.* Orlando, FL: Harcourt Brace.

Henderson, N. (2013). Havens of resilience: Resilience and learning. *Educational Leadership, 71*(1), 22–27.

Jordan, L. (2005). Science inquiry: Is there any other way? In R. H Audet & L. K. Jordan (Eds.), *Integrating inquiry across the curriculum.* Thousand Oaks, CA: Corwin Press.

Levine, P. (2003). The civic mission of schools. *National Civic Review, 92,* 63–65.

Liptrot, J. (2010). *K–12 service learning standards for quality practice evaluation tool.* Retrieved from http:/www.maryandpublicschools.org/NR/rdonlyres/CEFD2869-9129-46A3-91CE443928D1ED6C/24871/K12_ServiceLearningStandardsforQualityPractice.pdf

McLaughlin, M., Irby, M., & Langman, J. (1994). At-risk students and resilience: Factors contributing to academic success. *Clearing House, 67*(3), 137–140.

Melchior, A., & Balis, A. N. (2002). Impact of service learning on civic attitudes and behaviors of middle and high school youth: Findings from three national evaluations. In A. Furco & S. Billig (Eds.), *Service learning: The essence of pedagogy.* Greenwich, CT: Information Age.

National Youth Leadership Council (NYLC). (2008). K–12 *standards and indicators of high quality service learning.* St. Paul, MN: National Youth Leadership Council.

Newman, J. (2005). Talents and type IIIs: The effects of the Talents Unlimited model on creative productivity in gifted youngsters. *Roeper Review, 27,* 84–90.

Newman, J. L., & Dantzler, J. (2015). Fostering individual and school resilience: When students at risk move from receivers to givers. *Journal of Community Engagement and Scholarship, 8*(1), 80–89.

Renzulli, J., & Reis, S. (2014). *The schoolwide enrichment model.* Austin, TX: Prufrock Press.

Roehlkepartain, G. (2007). *Service learning with disadvantaged youth.* Scotts Valley, CA: LSA NSLC.

Rutter, M. (1984, March). Resilient children. *Psychology Today,* 57–65.

Schlichter, C. (1987). Thinking skills instruction for all classrooms. *Gifted Child Today, 10*(2), 3–8.

Werner, E. (1984). Resilient children. *Young Children, 1,* 68–72

Werner, E., & Smith, R. (1992). *Overcoming the odds: High-risk children from birth to adulthood.* New York, NY: Cornell University Press.

Youth Service America (YSA). (2013). *Semester of service toolkit.* Washington, DC: Youth Service America.

CHAPTER 2

# SERVICE LEARNING AND STEM LITERACY

## A Case Study of Speech Students and a Natural History Museum

**Sally Blomstrom**
***Embry-Riddle Aeronautical University***

**Barbara Hayford**
***Wayne State College***

**Lori Mumpower**
***Embry-Riddle Aeronautical University***

### ABSTRACT

We explored how literacy in science, technology, engineering, and mathematics (STEM) can be developed through service learning. A study was designed in which a service-learning project paired a natural history museum with an aeronautical university; specifically enrolled were university students in speech classes. College students fulfilled an assignment to create audio files about museum specimens for museum visitors. The goal was to increase interest in and provide an online outreach for the museum. The university

*Science and Service Learning*, pages 19–48

offers STEM-related majors, including engineering, aviation, physics, and homeland security. Most students had an interest in STEM, however many were not familiar with natural history so research was required to develop the scripts and subsequent audio files. The script needed to follow guidelines set forth by the director of the museum, who gave students feedback before they made the digital audio files. The results indicated that students increased their STEM content knowledge and demonstrated STEM literacy, however, STEM literacy varied between participants as well as between science, technology, and engineering elements.

STEM education and competency receive national attention due to declines in student performance in international rankings. The United States ranked near the middle of 65 countries and regions in science and math based on the 2014 Programme for International Student Assessment (PISA) tests (OECD, 2014). One goal of national education reform in the United States is to move the nation's standings on the PISA exam from the middle region to the top within a decade (NSB, 2012). In general, 12th-grade students have shown some improvement in National Assessment of Educational Progress (NAEP) mathematics scores from 2005 to 2013. Results were unchanged, however, from 2009 to 2013, with 2013 overall results indicating that 3% scored at the advanced level and 23% at the proficient level (NAEP, 2014.) Science scores were mixed during the period. Questions remain regarding which variables are most important in sustaining improvement in STEM retention (Maltese & Tai, 2011). Student performance on STEM measures has resulted in a clamor for improved STEM education. A foundational component of STEM education is STEM literacy, a by-product of, and an essential building block for STEM competency. Arguably, STEM learning requires STEM literacy, so literacy is an important aspect of STEM education (Krajcik & Sutherland, 2010) and has been explicitly linked to service-learning projects (Reynolds & Ahern-Dobson, 2010).

The concept of scientific literacy was introduced prior to the concept of STEM literacy. In the National Science Education Standards (NSES), the National Academy of the Sciences (NAS, 1996) defines scientific literacy as

> the knowledge and understanding of scientific concepts and processes required for personal decision making, participation in civic and cultural affairs, and economic productivity. It also includes specific types of abilities. In the National Science Education Standards, the content standards define scientific literacy. Scientific literacy means that a person can ask, find, or determine answers to questions derived from curiosity about everyday experiences. It means that a person has the ability to describe, explain, and predict natural phenomena. Scientific literacy entails being able to read with understanding articles about science in the popular press and to engage in social conversation about the validity of the conclusions. Scientific literacy implies that a person can identify scientific issues underlying national and local decisions and

> express positions that are scientifically and technologically informed. A literate citizen should be able to evaluate the quality of scientific information on the basis of its source and the methods used to generate it. Scientific literacy also implies the capacity to pose and evaluate arguments based on evidence and to apply conclusions from such arguments appropriately. (1996, p. 22)

From this definition we focused on a person "being able to read with understanding articles about science" and "be able to evaluate the quality of scientific information on the basis of its source." Scientific literacy was explored in depth by Holbrook and Rannikmae (2009) who cited various definitions and views.

> Enhancing scientific literacy through science education is: developing an ability to creatively utilize appropriate evidence-based scientific knowledge and skills, particularly with relevance for everyday life and a career, in solving personally challenging yet meaningful scientific problems as well as making responsible socio-scientific decisions. But it is necessary to recognize that enhancing scientific literacy is also dependent on the need to: Develop collective interaction skills, personal development and suitable communication approaches as well as the need to exhibit sound and persuasive reasoning in putting forward socio-scientific arguments. (p. 286)

Refining further, we focused on the need to develop "suitable communication approaches," which we interpreted as encoding audience-appropriate STEM messages based on the research conducted and the assignment guidelines. Feinstein (2011) proposed

> Science literate people are competent outsiders with respect to science: people who have learned to recognize the moments when science has some bearing on their needs and interests and to interact with sources of scientific expertise in ways that help them achieve their own goals. It follows from this definition that the pursuit of science literacy is not incidentally but fundamentally about identifying relevance: learning to see how science is or could be significant to the things you care about most. (p. 180)

We focused on two facets of this definition: people have the ability "interact with sources of scientific expertise" and that science literacy is "fundamentally about identifying relevance" (Feinstein, 2011 p. 180). Students in our project needed to identify relevant information from credible sources. Krajcik and Sutherland (2010) suggested five instructional features to support students' skills in reading, writing, and communicating about science. A conceptual approach to scientific literacy was presented by Laugksch (2000), who recognized different approaches to measuring scientific literacy. The approach we adopted based on the research reviewed above is most closely captured in the approach used by science educators to measure science content knowledge. For example, Norris and Phillips (2003) separated science literacy into a fundamental and a derived sense, both of which can be applied to STEM literacy. The fundamental sense includes, but is not limited to, reading and writing STEM text, tables, and graphics. The derived sense includes, but is not limited to, application of STEM knowledge. Students in this project read and wrote STEM texts. They were asked to apply the knowledge from the texts to robotics. The focus of this chapter directly required the fundamental sense and to a lesser extent the derived sense.

The National Research Council of the National Academies' publication on *Successful K–12 STEM Education* (National Research Council, 2011) offered a broad definition of STEM literacy: "the knowledge and understanding of scientific and mathematical concepts and processes required for personal decision making, participation in civic and cultural affairs, and economic productivity for all students," (p. 5).

We integrate the focal points retrieved from our review above with the definition of science literacy put forth by Norris and Phillips (2003) to develop the following definition of STEM literacy: a demonstrated ability to read STEM text, tables, and graphics with understanding, to evaluate the quality of the read information, to identify relevant information, and to incorporate that information in a written or oral communication. This definition may be applied and evaluated by having STEM information translated into encoded messages that must meet criteria established by a science expert, in this case the director of the natural history museum, and the messages must be decoded so that they are interesting to the audience of K–8 students. Our working definition does not include the STEM literacy of the K–8 students who will be visiting the museum and seeing and hearing the exhibits. Finally, we suggest that STEM literacy projects be designed to incorporate and integrate all four STEM disciplines if possible.

In order to investigate STEM literacy in this context, service learning was selected as the pedagogical approach because students make connections between what they are learning in class and practical applications of that knowledge. Students additionally realize many benefits from engaging in service-learning projects, including understanding course material better, enjoying learning, liking service, receiving a professional development benefit, and gaining skills, experience, and confidence in their abilities (Isaacson & Saperstein, 2005.)

Communication is a practical discipline readily suited for contributions to society through service (Applegate & Morreale, 1999). Service learning has increased in popularity in the field. Service learning applications vary depending on the content area and the project, however, there are commonalities. Campus Compact (2015) defines service learning thus: "Service-learning incorporates community work into the curriculum, giving students real-world learning experiences that enhance their academic learning while providing a tangible benefit for the community." In the present project, college students created audio files for face-to-face and virtual visitors to listen to and learn from at the natural history museum. The Corporation for National & Community Service (CNCS, 2015) adds,

> High quality service-learning experiences motivate students to attend, participate, understand and master the academic lessons they need to succeed. Better yet, they see their efforts making positive contributions to the place they live, leading them to enjoy learning, respect themselves and make good decisions about the future. (para. 5)

In this project students demonstrated their motivation and enthusiasm when meeting with the museum director. For the most part, they understood and complied with the instructions for the assignment. They commented about the museums they had visited while in K–8 and could see how the audio files could enhance the experience. So they seemed to enjoy the project.

The college students were, for the most part, Millennials. Service learning offers an approach closely aligned with documented characteristics of these learners. It is important to mention that our discussion generalizes about Millennials. Generalizations do not hold true in all instances but can provide some understanding of the larger group. The cohort born between 1980 and 2000 has been referred to as Millennials, the Net Generation, Generation Y, Generation Next, and the Digital Generation (Raines, 2010). Raines (2010) states Millennial

> students must be actively involved in their classes, not just passive recipients of knowledge imparted by their teachers; they must be academically challenged and motivated enough by what they are learning and how we are teaching to put forth their best effort; they must have a lot of interaction with their teachers; and they must have the support they need to succeed. (para. 11)

Howe and Strauss (2000) described Millennials as individuals who generally like group activity, perceive it is desirable to be smart, are intrigued by new technologies, and are interested in grades. The interest in grades extends to an interest in education. Patten and Fry (2015) indicated that of the 68.39 million Millennials in 2014, some 21% of adult males (ages 18–33) had earned a degree, and 34% had some college, compared with 27% of adult females who earned a degree and 37% who completed some college. These percentages are lower than those of older age groups as measured at the time the data was gathered, but are higher than those of older age groups when age is held constant.

Service learning is an approach that was opportune for Millennial students in this project. The students were motivated because their speeches (audio files) were evaluated by the museum director, and the audience was a real audience of museum visitors. The students understood and appreciated how their speeches would be used, that they would be helping K–8 students visiting the museum. They liked the technology, and many worked with the instructor to record the audio files. Successful completion of the assignment meant they contributed to a museum and made it a better experience for visitors.

Service learning's major contributions to students' academic development has been shown to be "on higher order thinking tasks, such as analytical essays and case-based assignments, [in which] students in service-learning sections consistently performed better than their peers" (Felten & Clayton, 2011, pp. 79–80). We considered creation of the audio files an assignment requiring higher-order thinking because the content had to be specific and technical and the messages were to be interesting to a young audience.

We wanted to assess cognitive development, so we chose from the definition offered by Yorio and Ye (2012), which identifies "management skill development; writing skills; problem-solving skills; critical-thinking skills; GPA; course performance" (p. 11). Service learning should enhance writing skills, critical thinking skills, and course performance. The written scripts, and more importantly the audio files and the reflections, could each be assessed as a form of STEM literacy.

Finally, STEM literacy may be achieved through transitioning from a home language (or common vernacular) to a school language to science language. This transition may create a border crossing leading to transformative experiences and learning (Kiely, 2005; Yore & Treagust, 2006). A model was created by Kiely to show that service learning is inherently transformative. We further suggest that service-learning projects, Service-learning is an approach that was opportune for Millennial students in this project. written and oral science communication, can be used to enhance STEM literacy, creating an inherently transformational learning experience for students.

Service learning has been used to enhance STEM learning through innovative ways in a variety of STEM-related courses and courses required for students pursuing STEM-related degrees (e.g., Blomstrom, 2010; Ghosh-Dastidar & Tsenova, 2012; Hamann & Drossman, 2006; Hayford, Blomstrom, & DeBoer, 2014; Kammler, Truong, VanNess, & McGowin,2012; Tedesco & Salazar, 2006).

## RESEARCH QUESTIONS

Our interest was in examining the relationship between service learning and evidence of STEM literacy. We created an assignment in which we expected students to be able to read and write text for science, technology, engineering, and math. Positing STEM literacy in the fundamental sense of reading and writing about STEM, we wondered if it could be enhanced by students' experiences in the service-learning project. To accomplish our proposition, we asked two questions:

1. Does participation in this service-learning project enhance student's knowledge of the science content (biomechanics and natural history); and
2. Does participation in this service-learning project enhance student's STEM literacy?

## PROJECT DESIGN

This study was designed to incorporate science (biology), technology (creation of audio files), engineering (application to robotics and aeronautics), and mathematics (e.g., relating information on rates). We designed this study to ensure that the community partner, an expert in biology, worked directly with the professor and the student participants in the assigned service-learning project. The project was designed so we could analyze (document, test) whether the project enhanced STEM literacy (this incorporates our analysis in the project design).

### Community Partner

The community partner was the A. Jewell Schock Natural History Museum, housed on the campus of Wayne State College in Wayne, Nebraska. The target audience was selected based on the most representative visitors to the museum. The museum annually reaches nearly 2,500 people in the region, with 680 people visiting as part of tour groups. About 78% of these tours involve K–8 students from area schools. Thus, the target audience is made up of K–8 students from the region, with the goal of increasing the number of tours for this group while also reaching K–8 students who cannot visit via the digital audio tours created through this service-learning project. The audience for the project was 3rd- to 8th-grade visitors. Most visitors to the museum are from northeastern Nebraska, northwestern Iowa, and southeastern South Dakota, with students from the town of Wayne composing the greatest proportion of visitors. The audience from Wayne may serve as representative of the region and was well suited for the project, because we presented a single effort. First, students in Wayne represent both rural and small urban populations, with only one elementary, one junior high, and one senior high school serving the city and its surrounding area. Results from the Nebraska state-administered tests in 2014 indicated that over two thirds of Wayne 8th-grade students met or exceeded science standards

(NeSA, 2014a). In general, Wayne 8th-grade students ranked in the middle third of the country in science scores (NAEP, 2011).

This service-learning project met the need of the museum and target audience by providing digital audio tours. Few opportunities exist for school-age children to participate in informal STEM education in the region. Students visiting the museum, either in person or virtually, can hear college students give presentations about individual specimens. The visitor is likely to hear more than one audio file. We have designed this service following a mixed-age model of college students with elementary students, in which the college students act as role models for the young audience. In this model the college students are perceived positively. The college students' interest in STEM is demonstrated through their enthusiasm, which affirms the value placed on STEM as well as relating the science content in an interesting manner. The audio tours are lively and interesting. Our hope is that this enthusiasm is passed on to the visitors who get the impression that science is fascinating and valued.

The partnership developed because the director of the museum and the faculty member collaborated on previous projects, including service learning, and both perceived that the project could be mutually beneficial. The opportunity arose because the faculty member had expertise in teaching about and evaluating delivery skills but lacked expertise in evaluating STEM content. The museum director, who also taught biology, recognized that students may know the science content but could have difficulty relating it, especially to an intelligent audience unfamiliar with the content.

## Participants

Table 2.1 indicates the number of students in each declared major. During the spring semester of 2014, some 72 students were enrolled in three sections of Speech. Six students either withdrew or failed the course. Of the remaining students, 52 wrote scripts and created audio tours on museum specimens. Of the 52 students, 21 were freshmen, 19 were sophomores, 8 were juniors, and 4 were seniors. By college, the enrollments were 28 in the College of Aviation, 15 in the College of Engineering, 6 in the College of Arts & Sciences, and 4 in the College of Business. Students who were enrolled in the course belonged to 17 majors, all involving STEM coursework (Table 2.1). These disciplines are markedly different from natural history, a facet of biology focused on life cycles, reproduction, and scientific names of species. Although the students were working in a STEM field unfamiliar to them, they were also working on biomechanics, so they also had something of some interest to them linking the project to their STEM disciplines. The students were studying STEM majors and had a high likelihood of

**TABLE 2.1 Majors Represented With Numbers of Students**

| Major | Students |
|---|---|
| Aeronautical Science | 13 |
| Aeronautics | 3 |
| Aerospace Engineering | 4 |
| Air Traffic Management | 2 |
| Aviation Maintenance Science (AS) | 2 |
| Aviation Maintenance Science (BS) | 4 |
| Business Administration | 4 |
| Computational Mathematic | 1 |
| Computer Engineering | 2 |
| Engineering Physics | 1 |
| Homeland Security | 3 |
| Human Factory Psychology | 1 |
| Mechanical Engineering | 5 |
| Space Physics | 1 |
| Still Exploring Engineering | 4 |
| Unmanned Aircraft Systems | 2 |

communicating STEM material to audiences in the future. The assignment gave them the chance to understand content required by a STEM expert in the field, to research using recommended credible sources, and to make the content understandable and interesting to an audience with little, if any, exposure to scientific terminology.

The assignment was given to the service learning students in stages. First the overall concept was defined and described. An overview of the community partner was presented along with information about the museum's need to expand outreach and increase interest. A list of the specimens with biomechanical designations and photos of the specimens was provided by the community partner. Biomechanical designations were based on types of locomotion and included three groups: walkers, flyers, and swimmers. Each student was assigned a specimen and asked to write what was known about the specimen assigned. Next, a prototype of the script was made available. The community partner provided an exemplar to better show how to develop the script into an effective short audio tour. (See Appendices A and B for the script format and exemplar.) The exemplar did not include every aspect of the information asked for in the assignment, allowing us to point out the differences and stress the requirements during class discussion. The exemplar was easy to understand, which was essential for the audio files. We expected the students to be able to code switch (Kiely, 2005), that is, to convert academic language to commonly understood language in their audio files. The museum director provided a bibliography of credible sources for students to use in their research. The exemplar was used to demonstrate appropriate content. It differed from what was expected in two important ways: no sources were cited and the exemplar was too short. These two differences were discussed to stress the importance of having good content (which the exemplar included), the requirement to cite sources, and to include sufficient information to meet the length requirement (which the exemplar did not). The community partner reviewed each student's script and provided individual feedback with suggestions. The final product of the audio tours was evaluated by the museum director to determine the quality of the content and efficacy in communicating natural history, biomechanics, and the related robotics of the specimens to the target audience. After submitting the audio files, students wrote reflective papers in response to prompts (see Appendix C).

The museum director selected audio files based on how well the file met the criteria. The selected digital tours were made available via the Internet on the museum's web page. At the end of the semester after the audio files were graded, students were asked to complete a consent form. This provided two options: consenting to give the museum permission to use the audio file or not consenting to allowing the file to be used. Consent was completely voluntary and did not affect the grade as those had already been assigned.

The timeline for the project included most of the semester. Students were informed about the project in January. During the last week of February the materials were made available for the students along with detailed, explicit instructions. Scripts of the presentations were due on March 12. Feedback was available for the students before the end of March. Students had until the middle of April to create mp3 files. Feedback for those files was available by the end of April.

The service-learning project was a required assignment for the Speech course. Out of 100 points for the course grade, 4 points were allocated to the script and 11 points to the audio presentation. The reflection paper counted for 5 points. Students were to report basic scientific information about the specimen, describe attributes and characteristics of the specimen, describe the biomechanics, and relate the biomechanics of each specimen to advances in robotics. The project pushed students beyond their comfort zone in two ways. First, they were to work with content that was not directly related to their chosen field of study. Second, they were to learn new vocabulary associated with this unfamiliar field then translate that language into lay terms for their intended audience.

**TABLE 2.2 Relationship of Assignment to STEM Disciplines**

| STEM Component | Assignment |
|---|---|
| Science | Natural history of specimen |
| Technology | Relate biomechanics to robotics |
| Engineering | Biomechanics of specimens |
| Mathematics | Compare and contrast rates of speed |

The service-learning project was designed so that the audio tours created by participants incorporated a selected aspect of each component of STEM (Table 2.2). Credible web sources and STEM journal articles were provided by the community partner to assist participants in finding the necessary reading information for their tours. They had to read and interpret this information, marking a transition from their home language (familiar terms, etc.) to a new scientific language (e.g., Kiely, 2005; Yore & Treagust, 2006). We previously observed students attempt to communicate STEM subject matter at the higher level of science language or vocabulary in a variety of situations. By choosing the higher-level language, they essentially mimicked what they read. Yet they were often unable to comprehend the meaning behind the words. Thus, participants were assigned the task of translating the concepts and vocabulary back into not only a home language but one aimed at a much younger audience from grades K–8.

Translation of the concepts for the audio tour necessitated deliberate decisions by the students about what to include in their short audio tours. This design element promoted cognitive development in the form of analytical and critical thinking forming the bases for content selection. Finally, the students were instructed to make the audio tour as interesting as possible with the aim of using the college students' enthusiasm to incentivize the young audience to explore more museum displays and more STEM topics in general.

## METHOD

We used a mixed-methods approach with a triangulation design to explore the research questions using qualitative and quantitative data sources (Borrego, Douglas, & Amelink, 2009). We had not found an instrument to measure STEM literacy in the literature suitable for our project. We concluded that the mixed-methods approach could shed light on the different data sources. In order to answer our research questions, we looked first at the qualitative data available in the scripts. The scripts provided information about students' content knowledge. The scripts were analyzed for indications of students' progression in terms of subject matter knowledge. The scripts were read with an eye to see if the requested content was included, and if so, whether it was accurate and sufficiently detailed. Students received feedback from the community partner, the museum director. We expected that students would incorporate feedback so the audio files would show improvement. The audio files were evaluated using a quantitative instrument. The items in Appendix A were each assigned a point value. A student's audio file was scored on each item for completeness and accuracy (a score of 1 point), a partial answer (a score of .5), or no answer/wrong answer (a score of 0). The evaluation tool corresponded directly to the assignment instructions.

Student responses to reflective prompts were qualitatively analyzed. The prompts were designed to answer one of the research questions (Table 2.3) or were related to service. The last prompt, for example, explained that the speech assignment addressed a real-world issue (evidence of low STEM knowledge and interest for K–8 students, particularly for those who would be visiting the museum), asking whether the assignment was effective and to explain why it was or was not effective. All reflections were read for content related to the first research question, and then all reflections were read for content related to the second research question.

**TABLE 2.3 Relationship of Reflection Prompts to Research Questions or Project Service**

| Reflection Prompt | Research Questions Addressed |
|---|---|
| Identify their assigned specimen. | Question 1 |
| Identify two discoveries they made. | Question 1 |
| Discuss what was done for the project including the research conducted | Question 2 |
| Identify how they incorporated effective informative speaking characteristics in their presentation | Question 2 |
| Write about their audience analysis and how the analysis affected their planning. | Question 2 |
| Identify what they would have done differently. | Question 2 |
| List their main points and write out what they knew at the end of the project about each of those points. | Questions 1 & 2 |
| Comment on how the presentation caused them to know the subject matter better. | Questions 1 & 2 |
| Explain how the experience helped them better understand what they were learning in the course. | Questions 1 & 2 |
| Discuss how that research would be similar to research they would do in their future. | Question 2 |
| Was the assignment effective service? Why or why not? | Service |

## RESULTS

Students' scripts provided formative information about the progression of their content knowledge. The scripts all indicated an increase in content knowledge based on responses students wrote when they were asked to write what they knew about their topics at the time the topics were assigned. Some of the scripts were quite limited, including only basic information about the specimen, but had little to no information on biomechanics and robotics (seeTable 2.4). Other scripts were better developed and more comprehensive (see Table 2.5).

The exemplar script in Table 2.4 has some interesting information, but does not meet all of the assignment requirements and is too short. Table 2.5 shows only an excerpt of a script. The exemplar script in Table 2.5 shows that the student went beyond the requirements to find and incorporate material from additional sources than those provided by the community partner. More complete information was provided on biomechanics and applications to robotics. Some scripts were more limited than the script shown in Table 2.4, and a few scripts were more comprehensive than the script shown in Table 2.5. These samples provide an illustrative indication of differences between the scripts.

The degree of content knowledge demonstrated in the scripts varied (see Tables 2.4 and 2.5). There are several possible explanations for why a large gap existed between basic and better developed content in the scripts. A few students indicated that the journal articles were hard to understand. Some students may have not taken the assignment seriously and consequently did not follow the format of the exemplar. After the scripts were reviewed, the instructor reinforced what was required in the assignment prior to the recording of the audio files.

The audio files were evaluated using a quantitative instrument. The results of the audio file quantitative analysis are shown in Table 2.6. The quantitative results indicated students overall included subject matter knowledge for the science (biology) portion, with a range of average student scores between 91% and 98% for the general information items on the specimen, with the exception being the specimen's functional role in nature with students scoring 14% on average. Students, on average, scored between 77% and 79% on the biomechanics (engineering) items. The comparison with other types or the rate of movement items (math) showed a combined average of 69%. The average score for application to robotics (technology) was 35%.

Scripts were submitted by 53 students. Of those students, 14 students received full points. Over 50% of the students turned in above-average scripts and digital audio tours. The students who included extra research went on to create very interesting audio files, demonstrating their STEM literacy in this project. Of the audio files, 16 were currently on the museum's web page at the time of this writing. Students demonstrated their ability to read, interpret, analyze, and critically select appropriate material and then

**TABLE 2.4 Example of Student Script for the Project Indicating Minimal Effort**

1. General information on the specimen
    a. This animal is the Didelphis virginiana, which is part of the phylum chordata
    b. More commonly known as the American or Virginia opossum
    b. The American or Virginia opossum is a mammal.
    c. They have four legs and an endoskeleton, which means that their bones are on the inside of their body. According to NatureWorks, the Virginia opossum is a marsupial, which means that they have a pouch sort of like a kangaroo. They also have a tail that is prehensile, which means they can grasp and wrap around things such as tree branches. Another interesting fact is that they have an opposable hallux, which is like a thumb. This also helps them grasp branches.

2. Habitat (cite source)
    a. The opossum lives in a wide variety of habitats like forests, open woods, and farmland.
    b. According to NatureWorks, they often prefer to live in wet areas such as marshes, swamps, and streams.

3. Diet (cite source)
    a. According to Leila Sicilian Martina, Animal Diversity web staff at the University of Michigan, opossums are omnivores, which means they eat both plants and animals. They usually eat vertebrates, invertebrates, plants, fruits, and grains.

4. Biomechanics
    a. The Oxford dictionary defines biomechanics as the study of the mechanical laws relating to the movement or structure of living organisms.
    b. The opossum walks on four legs and has multiple joints, which helps them move. They have an internal skeleton and their muscles connect to the bones, and that helps with their movements and helps them jump and climb.
    c. The opossum's movement is like many other four-legged animals, like lions, bears, and tigers.

5. Application of those biomechanics in robotics (cite source)
    a. This helps us build robots. We can use the idea of muscles connecting to bones to help our robots move. We use pistons, which are the equivalent of the muscles in the opossum, and that will help the robot move.
        http://www.nhptv.org/natureworks/opossum.htm
        http://animaldiversity.ummz.umich.edu/accounts/Didelphis_virginiana/

*Feedback for this script:*
Make sure to underline your genus and species names. I know that this is a rough draft, but you do not provide enough information for me with which I can provide guidance. You need to expand on how muscles work with the endoskeleton to produce movement. You will need to lengthen this considerably to meet the requisite time for the audio tour. Phylum Chordata is capitalized. I do like the information about how they use their tails.

communicate the information in lay terms. Their work carried an implied value statement about STEM. Interest in the specimens and in the project was conveyed through their enthusiasm and their attention to detail. Museum visitors who hear the audio files created by students who were engaged in the project may be more likely to retain the information, and they may have more interest in science or STEM themselves.

**TABLE 2.5 Excerpts From a Script Exemplifying Student Retrieval and Use of Informative Details on Biomechanics.**

1. General information on the specimen
   a. Hello and welcome to the Aphonopelma chaleodes exhibit.
   b. This specimen is better known as the tarantula.
   c. It belongs in the arachnid category and is reputable for being one of the largest of its type. Females can live up to twenty years of age while males only live up to around seven years of age (If not eaten by female after mating) (Szalay, 2014).

4. Biomechanics
   a. Biomechanics is the study of the mechanics of the movement of living organisms (biomechanics, 2003).
   b. Tarantulas traverse over terrain with eight legs that have multiple joints that lead up to the body. Their body is that of an exoskeleton and when the muscles inside the skeleton contracts, the leg moves. Arachnids use arthropod movement to stabilize like a tripod when the other legs are in the air. This allows very stable and rapid transportation over uneven surfaces (Arthropod Locomotion Engineering).
   c. This type of arthropod movement compares to other animals as a more overall stable method. Because the legs are jointed, these act as shock absorbers which help the tarantula glide along (Arthropod Locomotion Engineering).

5. Application of those biomechanics in robotics
   a. The application with biomechanics to be used in robotics is just at the embryonic stage. Large potential is seen with arthropod movement in the near future. The way the legs work for arthropods is giving scientists the ability to traverse uneven terrain very easily. These methods could be carried out in the ocean and even on different planets for exploration (Arthropod Locomotion Engineering).

*Work Cited*

Arthropod Locomotion Engineering. Retrieved March 14, 2014, from http://shapeoflife.org/video/other-topics/arthropod-locomotion-engineering

biomechanics. (n.d.) The American Heritage Dictionary of the English Language, Fourth Edition. (2003). Retrieved March 14, 2014, from http://www.thefreedictionary.com/biomechanics

Szalay. (2013 September 25). Tarantula Facts.LiveScience. Retrieved March 2014, from http://www.livescience.com/39963-tarantula.html

Tarantula Habits. (2014). Retrieved March 14, 2014, from http://en.wikipedia.org/wiki/Tarantula#Habits

*Feedback*

This is information is excellent. You may need to work at it to make the audio tour last the requisite amount of time, though. You could include the information about the Museum similar to how it is presented in the exemplar audio tour.

## REFLECTION ANALYSIS

The written reflections also were analyzed. We looked for evidence of research skills and encoding skills. These skills specifically were interpreted to be the skills to understand what was read and then write and deliver a message in language easily understood by a lay audience. We also looked for statements of applying the skills to other situations. All of the students

**TABLE 2.6 Spring 2014 Evaluations**

| Did you give an overview of the museum? | Points Possible | Average Points Earned | Total Points Earned | Average % of Possible Points |
|---|---|---|---|---|
| 1. General information on the specimen | | | | |
| Did you give both the scientific and common name? | 1 | 0.96 | 49 | 94% |
| Did you give the type of organism? | 1 | 0.99 | 50.5 | 97% |
| Did you give defining characters ? | 1 | 1.00 | 51.00 | 98% |
| 2. Habitat (cite source) | | | | |
| Did you define where they live? For example, bees live in hives. | 1 | 0.98 | 50.00 | 96% |
| Did you include geographic description of where they live(d)? | 1 | 0.84 | 43.00 | 83% |
| OR | | | | |
| Did you include geological time period in which they lived? | | 0.10 | 5.00 | 10% |
| Did you describe the specimen's diet? | 1 | 0.93 | 47.25 | 91% |
| Did you define their functional role in nature? | 0.25 | 0.14 | 7.10 | 14% |
| 3. Biomechanics | | | | |
| Did you define biomechanics? | 1 | 0.81 | 41.25 | 79% |
| Did you describe how they moved, their biomechanics of movement? | 1 | 0.79 | 40.15 | 77% |
| Did you describe how this type of movement compares to other types of the same movement? | 1 | 0.51 | 26.15 | 50% |
| And/or did you describe their rate of movement. | 0 | 0.20 | 10.00 | 19% |
| 4. Application of those biomechanics in robotics (cite source) | | | | |
| Did you show how the biomechanics of your specimen or closely related groups are used in robotics? | 0.5 | 0.35 | 18.10 | 35% |
| 5. Did you mention/cite appropriate sources? | 0.25 | 0.15 | 7.80 | 15% |

demonstrated increased scientific knowledge based on the quantitative analysis. The following are indications of some of the thoughts about research shared by the students. The reflective comments included in the remainder of this chapter may be (but probably are not) written by students whose scripts were cited previously. The analyses were conducted separately without knowledge of scores received on the quantitative measure.

Our definition for STEM literacy is a demonstrated ability to read STEM text, tables, and graphics with understanding, to evaluate the quality of the read information, to identify relevant information and incorporate that information in a written or oral communication. In examining the reflections for evidence of research, all students wrote about conducting research. Most had little knowledge of their assigned specimen prior to their research. We looked for the demonstrated ability to read STEM materials with understanding. Some students indicated the journal articles provided were difficult to understand. One student stated the following:

> Most of the work for this project was research. Biology has never been my best subject so having a list of sources was very helpful. However, a lot of them were very dense so I had to do outside research to understand what some of the articles were talking about.... My main points were the habitat of the Hawksbill Sea Turtle, its diet and its biomechanics. I learned a lot about each point because I had previously not known anything about the subject. I was able to gain information on these topics by researching via the sources that were provided. I think that presenting definitely helped me get to know the subject matter better.

We looked for indications of how students evaluated the quality of the information. Some students evaluated the quality by consulting multiple credible sources, as is evidenced by these reflective comments:

> I looked up general information of the beaver, specifically looking for its habitat, habits, physiology, and diet. I looked up other sources to confirm and cross check the info, and learned additional facts the others sources did not have.
>
> I researched everything on the opossum and filtered out the unimportant information. I made sure that the information was correct by looking at multiple sources and confirming the information I had.
>
> I thoroughly read over all of the informative documents that were provided. Alexander's piece was the most helpful because it included the most about mechanics. Unfortunately there was not enough information from the provided sources, so I searched for more. I ended up using five sources to get enough information. Other than the provided sources, I used credible sources like a BBC informative TV series called *Walking with Dinosaurs,* a biology textbook that had extensive content on the dinosaurs, and a publication by a professor from Stanford.
>
> In order for me to acquire information about the family, I had to use websites that were generally affiliated with museums and/or universities to ensure credibility because many other websites were not credible. With that, I researched the time period of which the Oreodont family existed and how they evolved over time in their natural habitat. This kind of research is beneficial because it built my skill of searching for credible online sites. I spent most

> of my time on this project doing all of the research. I used many different websites and essays to find the information valuable to my creature, giving me accurate data that I was able to share on the audio tour. The research that I did for this project helped me understand that certain sources are certainly more helpful and accurate than others may be.

Another aspect of literacy is being able to use language different from the language in the journal articles so the audience finds it easier to understand and retain the information. Students needed to understand the information they read then relate that information to the intended audience members, who were younger elementary students visiting the museum. Adapting content from journal articles for use in audio files intended for a general audience implies familiarity with the language and with the subject matter. The following are typical responses from the students about audience analysis:

> The museum presentation increased my understanding of the concepts we are learning in this course. Understanding who your audience is guides how the presentation is ultimately created. I do not want to put college-level content for an audience of elementary students because they would not understand it. I organized it according to the example given and delivered it with some excitement when warranted.

> Always think about who are going to be speaking to before you start planning so the audience will actually pay attention more when you speak. I attempted to not have words the everyday person might not know, so I kept it relevant in the respect to a child visiting a museum pressing buttons and listening to what is said about the animal in front of them.

> This assignment made me realize that audience analysis is crucial. If I hadn't cared, I would have used huge scientific words that they never would have understood. Content development is important because I needed to have the correct information to present. Organizing the speech is different because the information needed to be easily understood. Delivery is important because they have to be able to understand me when I'm speaking over the recording. And incorporating feedback is important because I need to be able to take criticism and correct information and problems.

> I was able to analyze my audience by what we talked about in class. Most of the museum attendees are smaller children or families. Because of this I tailored my speech to a younger audience. I tried to use simpler language that they would be able to understand. I also used analogies so that younger people could get an understanding of what I was talking about. I can apply this to future career communication because it is important to tailor how you speak to your audience.

Students involved in this project study STEM-related disciplines. Another component of interest was how the students would apply the skills they learned. Three students did not find the assignment to be effective. The other students found the experience to be effective, and several were very positive about it. Most students wrote about how they could apply the skills from the assignment. Their reflections indicated they had an awareness of how the experience could be applied in their future research projects and in their future communication. Here are some of the reflections on the topic of application.

> This experience required me to research a subject that I had no knowledge about. I think it will help me in the future to research a subject in a timely and concise manner. In addition, this experience helped me to speak accordingly to the audience's level of understanding.
>
> In my career communication I can now effectively break down a topic into critical main points and present the information in an effective manner utilizing technology and limited interaction with the audience.
>
> This assignment was a great experience. I was able to utilize the skills I have acquired in speech to communicate scientific information to spur the imagination of a younger generation. While doing so I have worked in a real-world environment as far as a team and the technology aspects.
>
> This project reflects many types of research I will have to do in the future, being a topic where I will have to figure general data about it and present in an appropriate manner. I had done research on my topic by utilizing credible sources to find information regarding each main point of my topic.

We also wanted to know about students' metacognitive awareness, that is, what are the students aware they learned from the course, particularly through this assignment? The following are a few indications of students' awareness of applying course concepts:

> This experience furthered what I believed the course was trying to teach in the beginning: to teach individuals to complete a wide range of speeches under a wide range of circumstances and still perform at a high standard. This project was first time exposure to analyzing an audience that was forever changing and would offer no interaction whatsoever. It was a refreshing chance to organize content in a developing manner. It gave a different medium of delivery to the audience by utilizing technology that would incorporate no feedback from the legitimate audience.
>
> I have most improved upon my research skills. This has granted me more confidence in my abilities to perform a designated task. I did notice myself taking my audience into account more. I thought it was quite fun making the audio. It definitely helped me stay interested in the project. It was an effective way of

relating to younger students who have minimal knowledge of dinosaurs. If the audio was done with some energy, the student would have walked away with a stronger desire to learn, which I think is paramount for education.

While researching and presenting this topic, two discoveries that I made were that studying up on an insect or creature that I have never of heard before can actually be very interesting and makes me want to start looking up other animals or creatures and how they work... This experience better helped me understand what I am learning in the course because I was able to learn many things about what I should be understand[ing] and learning in this course. In order to complete this speech I had to do accurate and precise research on the topic in order to develop interesting content on the matter, but then I had to write the speech well so that the audience understands what I am talking about, but still making it knowledgeable. In order for the audience to retain the information, it had to be organized so they could follow what I am speaking about. Delivering the speech helped me understand how speaking can be very powerful if done correctly, allowing myself to learn how to make my voice more effective. Feedback enables me to understand what I can improve on with the speech so that I can improve my skills

## DISCUSSION

This project has several limitations warranting mention, including the small sample size and a limited number of class sections. A larger sample is needed to make stronger claims about the efficacy of the method used. Another limitation is that the students were studying STEM-related fields so they likely had a stronger interest in STEM and perhaps had better-developed STEM literacy skills to start with. Results might have varied if students had been pursuing other fields. In this project the scripts and audio files were evaluated by one person, the museum director. At this point in time we do not yet have feedback from museum visitors. To better understand the effectiveness of the content of the audio files for different audiences, we should compare results from live presentations with audio presentations and possibly written with live and with recorded presentations. We evaluated the files primarily for content and not delivery. When surveying museum visitors, it may be worthwhile to inquire about their interest in the content, retention of information, and impressions of delivery.

Although service learning participants were already studying in STEM fields, we designed the project so students went beyond their comfort zones. The students were creating audio files in biology, whereas their interests were primarily in engineering, aviation, business, homeland security, human factors, and physics. We anticipated that linking biology and natural history (not their STEM focus) to engineering (their STEM focus) via biomechanics would engage the students, but found instead the students were somewhat reluctant to gather information on biomechanics and robotics. Thus, even the biomechanics provided dissonance in the project. Kiely (2005) showed that dissonance resulting from encouraging service learning students to push beyond their comfort zone led to a border crossing or transformative experience that increased learning. We designed the study around this concept but did not collect data to determine whether the border crossing enhanced STEM literacy.

The mp3 files created by students for this project indicated that they could comprehend the standards of STEM literacy presented to them and create messages consistent with those standards. The script guidelines, the exemplar, and samples made explicit what was expected of STEM-literate people in this situation. Students were motivated to do a good job on the project because they had a "real" audience consisting of the museum director and future museum visitors. Although they did not meet the museum director or any of the visitors to the museum, the students seemed comfortable with the technology that allowed them to reach a virtual audience.

## COMMUNITY PARTNER AND FACULTY MEMBER REFLECTIONS AND SUGGESTIONS

We present discussion and lessons learned first from the perspective of the museum director and then from the perspective of the faculty member.

### Museum Director Perspective

As Director of the Museum, I am both the community partner and a STEM educator. By working across disciplines as we have, this partnership has highlighted how important STEM literacy is to my efforts in both formal and informal STEM education and outreach. In particular, I learned that communication is not an afterthought of a given STEM project, but is necessary in guiding STEM learning. I was pleased to work with students on STEM subjects tangential to their own interests and to see new interests result from their service to my museum. Although we gathered no data, I did observe museum visitors react with great interest to both the basic biology and the biomechanics in the digital audio tours.

Review of a written script is an important and necessary part of guiding the students on the STEM content of their tours. Future studies should include review of the script in addition to the face-to-face feedback from the community partner. This results in more work for the community partner, thus I recommend that new partners have a frank discussion about the amount of effort necessary for a project such as the one described herein.

I also recommend that new partners undertake this extra work given the excellent results, which for the community partner included creation of valuable, unique, and interesting STEM outreach products.

We respect each other and work well together despite considerable geographic distance separating our two institutions. Our willingness to listen to each other and to change has allowed this project to grow and improve over time. My recommendations for moving forward are to continue the iterative nature of the project, to gather data from visitors to the museum, and to create new STEM communication products for direct delivery to schools.

## Faculty Member Perspective

As the faculty member, I deeply value this experience for my students because their speech is evaluated by a content expert who provided individual feedback. Additionally their speeches could have a broad audience of people they would never meet. The reach of their work far exceeded what is typical of students in most speech classes. Their work mattered. Visitors to the museum both in person and via the web would listen and gain information and, perhaps more importantly, might become interested in science, technology, engineering, or math. This is an exciting project and one I am proud to be a part of and for which I am grateful. I work well with our community partner in large part because we respect each other. We also respect and appreciate each other's areas of expertise. I love learning from my partner in this project. Today my knowledge of animals and their biomechanics and applications to robotics far exceeds what it was before we started this project.

Based on this experience, I learned we need to have technology considerations worked out on the front end. It turned out that institutionally we were not set up to give our community partner access to our learning management system. Consequently, I downloaded all of the files and posted them in Dropbox. That would have been fairly simple, but I had not specified a naming protocol, so I needed to rename each file when I saved it. A simple naming protocol such as animal name and section number would have facilitated the process. I also learned that instructions need to be very clear. I told students not to state their names in the audio files, but some did, and we could not use those files. I gave students instructions for creating mp3 files, but some created other types of audio files, and some of those did not work with the software the community partner and I were using. Now I record the speeches rather than giving students the choice of recording them on their own or coming to my office for me to record them. I also learned that providing more examples of audio files and scripts would help the students as they were preparing their presentations. In subsequent classes, I found it useful to have some examples of files that were not acceptable and discussed the characteristics that were positive about those files and identify the aspects that made them unacceptable. I encourage anyone who is considering a similar project to complete IRB training early and submit the required forms in a timely fashion. I also want readers to know that this is a fun project. Be prepared to be surprised. We found out that museum visitors really liked the diversity of voices, which was something we had not considered at the start.

We wrote about our initial experience here. Since that time, we have revised the project twice for two subsequent semesters. Face-to-face interactions with the museum director enhanced students' experiences and when students heard other students presenting to the museum director, they worked to improve the quality of their speeches. We continue to make refinements, and the project is continuing to expand. We learn each semester and find that participants in the classes as well as museum visitors enjoy the project. This service-learning project has improved my classes for my students and for me personally. The interactions and reflections clearly enhance my effectiveness and they make the project enjoyable as well as rewarding.

## APPENDIX A
## Script Format

1. General information on the specimen
   a. Scientific Name (underline genus and species names).
   b. Common Name
   c. What type of organism is this specimen? For example, does it belong to mammals or insects, or birds, or reptiles, or segmented worms, etc. (cite source)?
   d. What characteristics define this specimen? For example, if it is an insect then it will have an exoskeleton, six legs, and most likely one or two pairs of wings. If it is a mammal, it will have four walking legs (except for bats which have modified their front legs into wings), fur, and mammary glands. And if it is a bird, it will walk on two hind legs, have front legs modified as wings, and have feathers, etc. (cite source).
2. Habitat (cite source)
   a. Define where they live. For example, bees live in hives.
   b. Include geographic description of where they live(d) if possible(cite source).

OR

   b. Include geological time period in which they lived (if known) (cite source) (Wikipedia is a good source for this).
   c. Diet (cite source)
   d. Define their functional role in nature. For example, beavers are plant eaters but they also modify their environment through cutting down trees and building dams. For example, peregrine falcons are predators. What did they eat? For example, beavers eat trees. For example, peregrine falcons eat small prey, including rodents, smaller birds, etc.
3. Biomechanics
   a. Define biomechanics (cite source)
   b. Describe how they moved (cite source). For example, insects walk on six legs. Their legs have many joints. They have an external skeleton so the muscles attach on the inside of the hard external skeleton, and when the muscle contracts this moves the leg. Because they have so many joints they are very good at climbing over complex surfaces. Or you could describe flight in insects.
   c. Comparison (cite source). How does this type of movement compare to other types of the same movement? For example, how

does insect walking compare to human walking? How does insect flight compare to bird flight?

OR

c. Rate of movement and analogy (cite source).

4. Application of those biomechanics in robotics (cite source)
   a. Did you show how the biomechanics of your specimen or closely related groups are used in robotics?

## APPENDIX B
## Museum Audio Guide Exemplar Script

Welcome to the A. Jewell Schock Museum of Natural History at Wayne State College.

The mission of the Museum is to preserve the natural diversity of the region through maintaining and adding to the collections, through science education and outreach, and through actions to conserve and restore natural habitat in the region.

The museum collections include nearly 20,000 specimens of plants and animals.

Today I am going to take you on an audio tour describing the biomechanics of one of these specimens.

**General Background**

Paratendipes thermophilus is commonly called a blood worm.

Although it is called a worm, this specimen is actually an insect. It belongs to group of aquatic flies called nonbiting midges.

Blood worms are the immature stage of nonbiting midges. These immature stages live in the water. They form pupae, just like caterpillars do, and then the adult emerges from the pupa at the surface of the water where it flies away (Armitage et al., 1995).

**Characteristics**

It is called a blood worm because it is a dark red color. The red color comes from hemoglobin.

Here is an interesting fact: Some insects have hemoglobin like humans do, but the hemoglobin of blood worms is more complex. It needs to be complex because it carries oxygen to the blood worms when they live in nearly oxygen-free water.

Blood worms have three distinct body parts, just like adult insects: a head, a thorax, and an abdomen. However they have a softer body as their exoskeleton is not hardened. They have two small legs near their heads and two hind legs at the end of their bodies (Armitage et al., 1995).

**Habitat**

They burrow in the soft sediment at the bottom of hot springs and warm pools. These soft sediments have very low oxygen, so blood worms thrash about to increase the flow of oxygen past their bodies. They feed on small broken pieces of organic matter called detritus.

Paratendipes thermophilus lives in the west and southwestern United States (Hayford, 1998).

We don't have a good record for how long Paratendipes thermophilus has been on Earth, but in general, nonbiting midges have been around since the Mesozoic Era (Wikipedia, 2014).

**Biomechanics**

Studying biomechanics informs us of how species live in their environment.

Biomechanics is the study of the form and structure of living systems as affected by physical forces and displacements (Penn State University, 2014).

Blood worms move using a hydrostatic skeleton. Their body is one long tube filled with fluid. They use circular muscles to squeeze the front of their body forward and they use long muscles to pull the back of their body forward. Think of their bodies like a long water balloon. When you squeeze it with your hands you push it forward (Wikipedia, 2014).

Blood worms also whip side to side by contracting muscles on one side of their body relative to the other side (Armitage et al., 1995).

Unlike other "true worms," blood worms also may move about on the small legs located on their thorax and the hind end of their abdomen. These legs are relatively soft and not segmented but produce enough of a fulcrum off of which to propel their body.

When they become adults they will walk through use of their exoskeleton and will fly with their two wings.

**Robotics**

Adult Paratendipes thermophilus may serve as a model for robots like other insect do. Insects walk with three legs down and three legs up providing stability, speed, and efficiency of movement (Insect Biomechanics and Robotics, 2014). Stable, efficient movement will help robots navigate difficult terrain on Earth and other planets.

**Conclusion**

This concludes the audio tour on the biomechanics of Paratendipes thermophilus and blood worms. I hope you have enjoyed your tour and return to learn more about the Museum and its specimens.

Insect Biomechanics and Robotics: http://bnelson102690.wordpress.com/2013/11/18/insect-biomechanics-and-robotics/

## APPENDIX C
## Reflective Questions for Presentations Spring 2014

Please write out responses to these questions and submit them through Blackboard. This assignment is to be completed within one week of your audio presentation.

1. What was your topic?
2. What did you do for the project? Please specifically include the research you did for the project and discuss how the research you did may be similar to research you will do in the future.
3. List your three main points and write out what you now know about each of the points. How did you gain information about your topic? What did you learn about your topic through this project? Did presenting on the topic cause you to get to know the subject matter better?
4. Three characteristics of effective informative speaking have been identified: a speech should be intellectually stimulating, relevant to the audience, and creative. How did you incorporate these three characteristics in your presentation?
5. Identify two discoveries you made. What did you learn about yourself? Did you employ a new creative skill? Do you see yourself as more confident? Did you become aware of assumptions you held—perhaps about the audience? Did you learn something new about delivering a presentation using technology?
6. How did you analyze your audience? How did your analysis affect your planning for your presentation? How can you apply what you learned about adapting to the audience in your future career communication?
7. In five lines of text, describe what you would do differently. For example, would you prepare introductory remarks differently?
8. How did the experience better help you understand what you are learning in the course? Please address audience analysis, content development, organizing your speech, delivery, and incorporating feedback.
9. Please describe how you can use what you learned from this experience in your career communication.
10. Did you come across the way you wanted to? What elements do you want to work on in the future?
11. You learned your personality profile. What new information did you learn about your personality type? How did having that knowledge about yourself influence your effectiveness in the group? How can you use that information in your career communication? How might having the information about your personality preferences

influence decisions you will make regarding your career such as determining which, if any, minor to pursue, which types of jobs you think you are best suited for, and what you recognize as not being a good fit for you?

12. This speech assignment has the elements of students addressing a real-world issue (STEM knowledge and interest for K–8 students visiting the museum), working in teams to give feedback, and delivery a presentation using technology. Was this assignment effective for you? Why or why not?

## REFERENCES

Applegate, J. L., & Morreale, S. P. (1999). Service-learning in communication: A natural partnership. In D. Droge & B. Ortega Murphy (Eds.), *Voices of strong democracy: Concepts and models for service-learning in communication studies.* Sterling, VA: Stylus.

Blomstrom, S. (2010). Identifying teaching effectiveness: Using student skill surveys, speech evaluations, and quiz scores to inform instruction. *Communication and Theater Association of Minnesota, 37,* 116–132.

Borrego, M., Douglas, E. P., & Amelink, C. T. (2009). Quantitative, qualitative, and mixed research methods in engineering education. *Journal of Engineering Education, 98*(1), 53–66. Retrieved from https://www.cs.auckland.ac.nz/courses/compsci747s2c/lectures/borrego.ea-quantitative-jee-09.pdf

Campus Compact. (2015). *Service-learning.* Retrieved from http://compact.org/initiatives/service-learning/

Corporation for National & Community Service (CNCS). (2015). Learn and serve America fact sheet. Retrieved from http://www.nationalservice.gov/newsroom/marketing/fact-sheets/learn-and-serve-america

Feinstein, N. (2011). Salvaging science literacy. *Science Education, 95,* 168–185. doi:10.1002/sce.20414

Felten, P., & Clayton, P. (2011, December 8) Service learning. *New Directions in Teaching and Learning. 128,* 75–84.

Ghosh-Dastidar, U., & Tsenova, L. (2012). Bio-math mapping: Water quality analysis of Hudson River and Gowanus Canal: A SENCER-based summer project. *Science Education and Civic Engagement, 4*(1), 55–65.

Hamann, H. B., & Drossman, H. (2006). Integrating watershed management with learning: The role of information transfer in linking educators and students with community watershed partners. *Comparative Technology Transfer and Society, 4*(3), 305–337.

Hayford, B., Blomstrom, S., & DeBoer, B. (2014). STEM and service-learning: Does service learning increase STEM literacy? *The International Journal of Research on Service-Learning and Community Engagement, 2*(1), 32–43.

Holbrook, J., & Rannikmae, M. (2009). The meaning of scientific literacy. *International Journal of Environmental and Science Education, 4*(3), 275–288.

Howe, N., & Strauss, W. (2000). Millennials rising: The next great generation. New York, NY: Vintage.

Isaacson, R., & Saperstein, J. (2005). The art and strategy of service-learning presentations. Belmont, CA: Wadsworth.

Kammler, D. C., Truong, T. M., VanNess, G., & McGowin, A. E. (2012). A service-learning project in chemistry: Environmental monitoring of a nature preserve. *Journal of Chemical Education, 89*(11), 1384–1389.

Kiely, R. (2005). A transformative learning model for service-learning: A longitudinal case study. *Michigan Journal of Community Service Learning, 12,* 5–22.

Krajcik, J. S., & Sutherland, L. M. (2010). Supporting students in developing literacy in science. *Science, 328*(5977), 456–459. doi:http://dx.doi.org/10.1126/science.1182593

Laugksch, R. (2000). Scientific literacy: A conceptual overview. *Science Education. 84,* 71–94.

Maltese, A. V., & Tai, R. H. (2011). Pipeline persistence: Examining the association of educational experiences with earned degrees in STEM among US students. *ScienceEducation, 95*(5), 877–907. doi:10.1002/sce.20441

National Academy of the Sciences (NAS). (2015). *National Science Education Standards.* Washington, DC: National Academies Press.

National Assessment of Educational Progress (NAEP). (2014). Average scale scores for science, grade 8 by all students [TOTAL], year and jurisdiction: 2011. *NCES.* Retrieved from http://nces.ed.gov/nationsreportcard/naepdata/remoterepgen.aspx?app=NRCNDE&p=2-SCI-3-20113-SRPUV-TOTAL-NT-MN_MN-Y_J-0-0-8

National Research Council (NRC). (1996). *National science education standards.* Washington, DC: National Academies Press.

National Research Council (U.S.). Committee on Highly Successful Schools or Programs for K–12 STEM Education. (2011). *Successful K–12 STEM education: Identifying effective approaches in science, technology, engineering, and mathematics.* Washington, DC: National Academies Press.

National Science Board (NSB). (2012). *Science and engineering indicators 2012.* Arlington VA: National Science Foundation (NSB 12-01).

Nebraska State Accountability (NeSA). (2014a). Nebraska state accountability: NeSA [science] scores: Percent proficient: Meets/exceeds combined results for all grades tested. *Nebraska Department of Education.* Retrieved from https://reportcard.education.ne.gov/20132014/pg_NesaPerfOverall.aspx?AgencyID=00-0000-000

Norris, S. P., & Phillips, L. M. (2003). How literacy in its fundamental sense is central to scientific literacy. *Science Education, 87,* 224–240.

Organisation for Economic Co-operation and Development (OECD). (2014, February). *PISA 2012 results: What students know and can do: Student performance in mathematics, reading and science.* Retrieved from https://www.oecd.org/pisa/keyfindings/pisa-2012-results-volume-i.htm

Patten, E., & Fry, R. (2015). *How Millennials today compare with their grandparents 50 years ago.* Retrieved from http://www.pewresearch.org/fact-tank/2015/03/19/how-millennials-compare-with-their-grandparents/#!13

Raines, C. (2010). *Millennials at work.* Retrieved from http://www.generationsatwork.com/millennials-at-work/

Reynolds, J. A., & Ahern-Dodson, J. (2010, July 1). Promoting science literacy through research service-learning—An emerging pedagogy with significant benefits for students, faculty, universities, and communities. *Journal of College Science Teaching, 39,* 24–29.

Tedesco, L. P., & Salazar, K. A. (2006). Using environmental service-learning in an urban environment to address water quality issues. *Journal of Geoscience Education, 54,* 123–132.

Yore, L. D., & Treagust, D. F. (2006, February). Current realities and future possibilities: Language and science literacy—empowering research and informing instruction. *International Journal of Science Education, 28*(2/3), 291–314.

Yorio, P. L., & Ye, F. (2012). A meta-analysis on the effects of service-learning on the social, personal, and cognitive outcomes of learning. Academy of Management Learning & Education, 11(1), 9–27.

CHAPTER 3

# GREEN WORKS IN SERVICE LEARNING AND STEM WORKFORCE DEVELOPMENT FOR URBAN YOUTH

**Alexis Petri**
***University of Missouri–Kansas City***

**Kate Corwin**
***Green Works in Kansas City***

## ABSTRACT

This study investigated the effectiveness of informal science learning opportunities for underserved youth. Green Works, in Kansas City, Missouri, was designed firstly as a phase-two program evaluation using an initial evaluation as a foundation and then, secondly, using Fuzzy cognitive mapping (FCM) as a process to elicit from youth participants how Green Works helps them meet their goals. Anecdotal evidence of Green Works' success exists for the youth and for their families and neighborhoods. Through student interviews using FCM as a research methodology, we identify promising practices that have led to desired outcomes for the urban youth and their community. Green Works educated urban high school students about Kansas City's environmen-

*Science and Service Learning,* pages 49–72

> tal challenges, involved them in service learning projects that improved their community, prepared them for postsecondary education and the STEM workforce, and connected them to opportunities in the growing green economy. The service learning best practices included involving STEM professionals, locating a majority of the service learning projects in the students' communities, choosing significant projects that tied directly to the curriculum and city priorities, and finding ways for the students to reflect, to write on the project, and to share the results with their friends and family.

Mario is one of many success stories behind Green Works, an innovative program combining service learning, environmental stewardship, career planning, and life skills in a year-round program over 2 years for inner-city youth in Kansas City, Missouri. One of the first students to complete the full Green Works program in Kansas City, Mario now has a green collar job at a water treatment plant (Corwin, 2014). He has attained two wastewater licenses (Corwin, 2015a) and gives back by serving on the Green Works board (Corwin, 2015b). Mario's career path was neither seamless nor easy. He grew up in a family that could be characterized as low income, graduating from an alternative high school. Immediately after high school, Mario matriculated at a local community college but attended less than a year. He found a job at a car dealership. Mario could be any young adult who struggles with science learning in formal academic settings such as high school and college. He was, however, successful in Green Works' informal learning environment, which gave him a strong foundation in self-confidence, social capital, and a dash of insight into the process of building a career. During his time with Green Works, Mario successfully completed internships with Kansas City Water Services, Habitat ReStore, and Kansas City Missouri Parks and Recreation (Corwin, 2015a). As a result of his successful internship, Mario gained employment at a water treatment plant (Corwin, 2015a).

## GREEN WORKS OF KANSAS CITY

This investigation was the product of a unique collaboration between two professionals who came together through a shared concern over urban education in Kansas City and the barriers facing the inner city's high school–aged youth. The perceived barriers included inadequate education, entrenched poverty, and a lack of opportunity. One author was a university service learning program director and applied researcher; the co-author was the CEO, founder, and sole full-time staff member of the nonprofit organization, Green Works in Kansas City. In this chapter, we demonstrate the importance of supporting the education available to teens living in Kansas City's urban core with informal learning opportunities that fill needed gaps in STEM education, hands-on learning, work experience, and social skills development.

Green Works began in the fall of 2007 with the mission of preparing urban teens for the workforce using an environmental service learning curriculum. Green Works provides environmental education, workforce development, and helps students gain social capital through partnerships with local businesses and organizations (Corwin, 2015). Several types of learning opportunities outside of school time are involved including science learning, environmental stewardship, job-skill training, internship experience, financial literacy, and life skills (Corwin, 2015b). The ultimate goal of Green Works is for program participants to grow into independent, resilient, and accountable adults actively engaged in their community (Corwin, 2015c). Green Works measures its success through increased science literacy, changed behaviors toward the environment, improvement in workplace skills, increased self-confidence, meaningful work experience, social-capital connections, and high school graduation (Corwin, 2014). Service learning is one of Green Works' main instructional strategies (Corwin, 2015b).

Green Works commits a minimum of 2 years of programming to each student accepted into its Year 1 program. Many students participate throughout high school and during the first part of their college career.

## GREEN WORKS: YEAR 1

Each year, Green Works accepts high school freshmen and sophomores into the Saturday program, Environmental Connection Opportunities for Students (ECOS), which meets for 8 hours each month. This full-year program combines hands-on environmental education, field trips, science labs, and service learning. Kansas City's most serious environmental problems are addressed, including water pollution, outdated sewer and storm water systems, low recycling rates, the substantial opportunity for weatherization of homes and office buildings, and a sprawling metropolitan area lacking a robust public transportation system. Seventeen organizations take part in ECOS by sponsoring field trips, helping with student experiments, and even leading students' lessons (Corwin, 2015d).

Students completing ECOS are invited to participate in one of two summer programs. Startup Summer introduces teens to numerous ways to make money without a job (i.e., offering pet care, computer services, organic lawn care, or maintaining flower beds). Startup Summer concludes with the students learning to manufacture and market an herbal insect repellent. Girls LEAD is a program that provides young women the opportunity to work with five environmental partners to learn the process of environmental auditing. Through this service learning program, the girls audit female-owned businesses and present their findings (Corwin, 2015d).

## GREEN WORKS: YEAR 2 AND BEYOND

Students participated in Green Future, meeting in the evening for 4 hours each month. Students completed interest and skill evaluations, practiced setting priorities and managing their time, created a master résumé, researched careers, and created three different career plans. Two of the plans that students created were for careers growing in Kansas City, paying at least $30,000 a year, and requiring two years or less of postsecondary education. Students also had opportunities for job shadowing and to practice interviews. At the completion of Green Future, students benefited from 8 hours of work readiness and soft-skills training. During the final summer of the 2-year program, students participated in internships at environmental organizations aligned with the students' career goals, interests, skills, and abilities, as demonstrated across the 2 years. Green Works paid the students stipends for participating in the internship (Corwin, 2015d).

Green Works continues to support the students' interest and engagement in environmental stewardship both during and after the 2 years of formal programming through Green Ink. All students were encouraged to participate in this service learning, community education, and advocacy program focused on the environmental movement. Green Ink met twice a month during the school year. Students undertook complex service learning projects such as the Blue River Rescue, creating a litter index, and making informational videos about recycling. They also tested water quality, conducted bird census counts, planted native trees and shrubs, and participated in community meetings involving environmental priorities such as public transportation (Corwin, 2015d).

## GREEN WORKS STUDENTS

To date, Green Works has taught 140 urban high school students about Kansas City's most serious environmental issues (Corwin, 2015c). The Green Works students live within the urban area of Kansas City public schools. The Kansas City school district is unaccredited. Of students, 28% scored proficient or advanced on the 2013 annual performance tests. The four-year high school graduation rate was 67% in 2013, up from 51% in 2011. The 2014 two-year average attendance rate was 71% (MDESE, 2014). Green Works has much different data. Of the 140 students served, 100% graduated from high school, and many continued their education in college. One recent graduate had the highest senior science score at her high school. Green Works students are leading by example: Not only are they reaching their goals and becoming better stewards of the environment in the process, they are also influencing their friends and family to take better care of the earth by recycling, conserving water and energy, and using natural products.

In 2016, some 39 students and alumni participatied in Green Works. The average age of a Green Works' student was 16.7 years. Of the student participants, 72% ($n = 28$) were female while 28% ($n = 11$) were male. In terms of demographics, 36 students identified as African American, 2 as Latina/Latino, and 1 as White; 41% belonged to low-income families that earn less than $27,000 annually and another 41% belonged to families that earned a moderate yearly income of $27,000 to $43,000. High income was set at above $43,000/year, still below the median income ($45,150) for Kansas City (U.S. Census Bureau, 2012).

One of Green Works' students, Ronald, graduated from a Kansas City public high school in 2013 and completed his second year of college. Ronald planned to earn a bachelor's degree and continue to law school with the intention of one day practicing civil law. Even though he was not planning on a green collar career, Ronald embraced the STEM learning nurtured by Green Works, cared deeply about the environment, and mentors Green Works' high school students through the program (Corwin, 2015a).

## Literature Review

Service learning is a viable means of responding to the growing demand for practical experiences that are more than altruistic that cross cultural boundaries and enrich all participants. According to the 2006 report, "The Silent Epidemic: Perspectives of High School Dropouts," more than 80% of students surveyed noted that academic failure alone does not lead to dropping out of high school. Of the surveyed students, 50% believed people drop out from high school due to disengaging coursework, while 70% of students reported a perceived lack of real-world applications as reasons why some youth abandon their education (Bridgeland, DiIulio, & Morison, 2006; Bridgeland, DiIulio, & Wulsin, 2008). Over 80% of the students in the survey believed incorporating service learning, and, similarly, work studies and internships, into high schools would have a positive impact on high school completion (Bridgeland et al., 2006; Bridgeland et al., 2008). Surveys of 807 high school students, including 151 students designated as at-risk pertaining to service learning, found more than 75% of students agreed that service learning classes are more interesting than traditional classes. Similarly, 80% of designated at-risk students not enrolled in service learning classes indicated that their attitude toward high school would likely improve if they were given the opportunity to participate in service learning (Bridgeland et al., 2008). Of the designated at-risk students who did not participate in service learning, 66% believed that service learning would motivate them to work hard (Bridgeland et al., 2008). According to the Corporation for National and Community Service (2006), students engaged in service learning reported confidence in their own ability to make a difference in their communities. Additionally, high school students participating in service learning described feeling good about giving back to their communities and about gaining insights into how the social world works (Billig, 2010; Root & Billig, 2008).

In a controlled study, Yamauchi, Billig, Meyer, and Hofschire (2006) found that service learning bolstered career exploration and expanded personal networks. Pre- and posttest results suggested that students who participated in service learning demonstrated stronger job skills and more explicit career aspirations than students who did not participate in service learning, according to Yamauchi et al., noted that these job skills included planning activities, engaging in effective communication, and demonstrating competence during job interviews. The "Growing to Greatness" report (Kielsmeier, Root, Pernu, & Lyngstad, 2010) called for students in low-income schools to be offered the same opportunities for service learning as students in high-income schools. The report identified partnerships with informal education and youth development initiatives as a route to offering all students access to the benefits of high-quality service learning. The goal for service learning is often to inspire a new generation of individuals with the civic and social interests and skills our democracy needs in order to thrive and to provide students with real-world experiences that open their eyes and minds to career paths.

## Workforce Development

Green collar jobs are those jobs that sustain environmental quality or perhaps even improve it (Falxa-Raymond, Svendsen, & Campbell, 2013; Kouri & Clarke, 2012), are in high demand (Arnett, Kitchel, & Cannon, 2011), and provide a living wage (White & Walsh, 2008). Two workforce needs have emerged: Green industries are rapidly growing and require a high-tech, STEM-capable workforce (e.g., solar power, wind power, biomass fuels). The greening of existing jobs also requires particular green collar expertise (i.e., an architect specializing in energy efficient materials) (Arnett et al., 2011). In 2006, more than eight million United States residents worked in green collar jobs, with estimates that jobs in this sector will grow to 25% of the U.S. workforce by 2030 (Arnett et al., 2011). This is good news for the country's struggling job market. Green collar job-training programs may be an important factor in lifting individuals and families out of poverty (Falxa-Raymond et al., 2013; Pinderhughes, 2007). Workforce development strategies for youth with low socioeconomic status are needed to help young adults gain the skills to build careers in the STEM fields (Falxa-Raymond et al., 2013). Successful workforce development connects students with specific employers for early career experiences and supports their transition to employment (Cooney, 2010; Hamilton, 2002). Strategic ways to support the link between young adults and business include building advisory boards with members of industry, allowing portions of the service learning curriculum to be designed by businesses, and creating opportunities for internship collaborations (Falxa-Raymond et al., 2013). Service learning can be a useful teaching methodology for helping high school students understand and prepare for green collar jobs. These types of initiatives are necessary to minimize the gap between the workforce need for green collar employees and the education available (Arnett et al., 2011). High schools need to include sustainability into their curriculum and consider their role in educating a green collar workforce (Arnett et al., 2011; Konopnicki, 2009).

## FUZZY COGNITIVE MAPPING

Fuzzy cognitive mapping is a method of making qualitative causal cognitive maps computable, which is gaining prominence in the fields of engineering, business, and information technology (A. J. Jetter, personal communication, November 13, 2013). In 1986, FCM was first conceived as a manner of compiling expert knowledge in a network of causal links (Kosko, 1986). Fuzzy cognitive maps build on qualitative cognitive maps by making them computable through the application of fuzzy set theory (Jetter & Schweinfort, 2010). One primary advantage of FCM is the structure it provides to make capturing expert knowledge an intuitive process. Through FCM, Green Works wanted to develop a knowledge base to help improve and grow the program. It was important that this knowledge base reflect the realities and perspectives of the students. The FCM works through a network of causal links and has been found to be a reliable knowledge-based model facilitating "sense making" by helping program partners and stakeholders communicate about strategies and decisions (Jetter, 2006). The first step in FCM is to gather qualitative data from key informants in order to construct cognitive maps. Then the key informants apply directional and weighted numerical values quantifying the relationships between major and minor factors. With the directional and weighted relationships added to the cognitive maps, the maps from multiple responders can be integrated to show a dynamic display of direct and indirect effects. With the weight and direction to the relationships between major and minor factors computed, the values can be adjusted to provide if/then scenarios for generating hypotheses and simulated testing (A. J. Jetter, personal communication, November 13, 2013). From a science education perspective, FCM claims to capture students' naïve understandings of the natural world and highlight how Green Works programs may or may not create important opportunities for implicit learning (Cole & Persichitte, 2000). For science learning to produce implicit knowledge available to the learner, a cognitive task is required that is substantially different from free recall tasks such as a focus group (Berry & Dienes, 1993, as cited in Cole & Persichitte, 2000).

## Method

Early in its programming, Green Works undertook a thorough evaluation in order to better understand the effects of the program (Campbell, 2009). According to the program evaluation, students who participated in Green Works demonstrated mastery of academic environmental science concepts, increased self-perceptions, confidence and concern about the environment, and an increased impact on the environment through example and influence (Campbell, 2009). Since 2009, Green Works has continued to grow its programming and make alterations based on student outcomes. While much of the program emphasis is on environmental stewardship, Green Works provides students a broad foundation supporting their transition to adult life. Green Works promotes high school graduation, personal financial management, postsecondary education, and early work experiences. Although such evaluation conclusions were formed from program evaluation, there was still a need to discover how the role of service learning in Green Works helped students meet their goals.

The authors identified FCM as a tool to help Green Works gain insight into the conceptual and causal relationships students attributed to its programs and the level of attainment of their goals (Cole & Persichitte, 2000). The most important factor in deciding to use FCM was our surmise that it would be empowering for the students because it establishes the students in their own lives as experts who are sharing their insights and perspectives to strengthen the program. We also wanted to expose students to a participatory evaluation process as part of providing exposure to basic tenets of business and assessment. This idea supported an important ethos running throughout the Green Works program and fostered increased self-confidence. The act of identifying and drawing the conceptual and causal linkages between Green Works and students' own goals had the potential of being much more representative of the students' actual perceptions than they might describe during either an interview or a focus group, or indicate on a survey.

The evaluation was designed to address two research questions. The first question focused specifically on the participants of Green Works—the urban high school students—and how the program helped them reach their goals. The second question focused on the outcomes of service learning.

1. How does Green Works help urban teen participants reach their goals?
2. How integral is service learning to the overall Green Works program experience?

## SAMPLE

The participants in the FCM project were students in Green Works who had completed at least one year of programming as of 2014. Green Works was serving 39 urban high school students. Maximum variation sampling (Creswell, 2007) selected participants with different experiences with Green Works programs and from different cohorts. Snowball/chain sampling identified additional participants who were "information rich" (Creswell, 2007, p. 127) and had participated in several years of Green Works programming. In 2014, there were 39 students participating in Green Works, and 14 of them (36%) participated in the evaluation by making maps. According to Jetter's (2006) framework, FCM has been studied with 7 to 46 individual cognitive maps.

## PROCEDURE

The Green Works FCM evaluation project followed Jetter's (2006) framework. This section presents a description of each step and its application.

### Clarification of Project Objectives And Information Needs

We, the evaluator and the CEO of Green Works, met to clarify project objectives and identify the evaluation questions.

### Plans for Knowledge Elicitation

Before beginning data collection, the knowledge sources and information gathering techniques were reviewed. In this evaluation of Green Works programs, the sources of the knowledge were the student participants. This information led to a single stakeholder design where one knowledge source authored one map.

## Data Collection

In order to better understand student perceptions about Green Works, students were invited to participate in the FCM project. The FCM project consisted of Green Works students meeting in small groups and also one-on-one to learn the fundamentals of creating a fuzzy cognitive map, understanding the difference between crisp logic and fuzzy logic, and stepping up as experts in their own lives. Following the explanation, participants completed a questionnaire and then began making their individual maps. A handout of "start concepts" was used for knowledge activation. The start concepts consisted of common beginning concepts, such as the Green Works programs and strategies. Students understood they did not need to use all or any of the start concepts.

The students were asked to draw a cognitive map depicting how Green Works was helping them reach a goal. They began by defining their selected goal (e.g., graduating from high school, going to college, building a career) and writing it on a large sheet of paper. The students were asked to place Green Works as a starting point on the opposite end of the paper. They were then directed to think about what programs, activities, and resources they gained from Green Works and to demonstrate on their map how Green Works helped them meet their selected goal. To their maps, the students added multiple concepts, connected the concepts with arrows, then weighted the strength of the connection. The students could rely on the "start concepts" list or identify their own concepts. Most students used a combination of the "start concepts" as a launching point then defined their own concepts as they neared their goal statement. The students did not write their names on their maps or questionnaires, but wrote a word of their own choosing on each so the maps and questionnaires could be matched. The questionnaires were designed to support and clarify the dynamic behavior of the system as depicted by each student participant. This approach helped to standardize some of the concepts in the maps.

# DATA ANALYSIS

The cognitive maps students drew were reviewed and clarified. The maps were combined by plotting each of the points on a matrix in Excel. This process made it possible to see patterns and common ideas as well as individual and unique ideas. After plotting each concept in Excel, the concepts were condensed into like groups. From there a matrix was constructed to show how the factors were related across all of the maps. Students completed 14 maps and 14 questionnaires. Each participant in this study introduced between 15 and 25 concepts. The concepts from the maps were analyzed and grouped into 42 concepts. The maps were then redrawn using *Mental Modeler*

(Gray, 2016), and the matrices for each map were combined mathematically. This process required analysis of each concept to combine as many related or overlapping concepts into one concept as possible. During the conceptual design process, concepts exogenous to the scope of the project were deleted.

## Mathematically Integrate Maps Into a Detailed Model

The goal of an FCM model is not to arrive at a definitive truth but to create a description adequate for examining the problem (Jetter, 2006). A detailed and useful FCM model should explain and predict how changes to the model or policy interventions impact the program. A useful FCM model also needs to be transparent to facilitate acceptance and implementation of the recommendations (Jetter, 2006). Our model sought to incorporate as many dimensions of the students' individual maps as possible, maintaining their voice and perspective. In this step of analysis, the individual maps were mathematically integrated into a detailed model. Figure 3.1 shows a model of one of the student maps, re-created by computer. Figure 3.2 shows the complex combined map and network of connections.

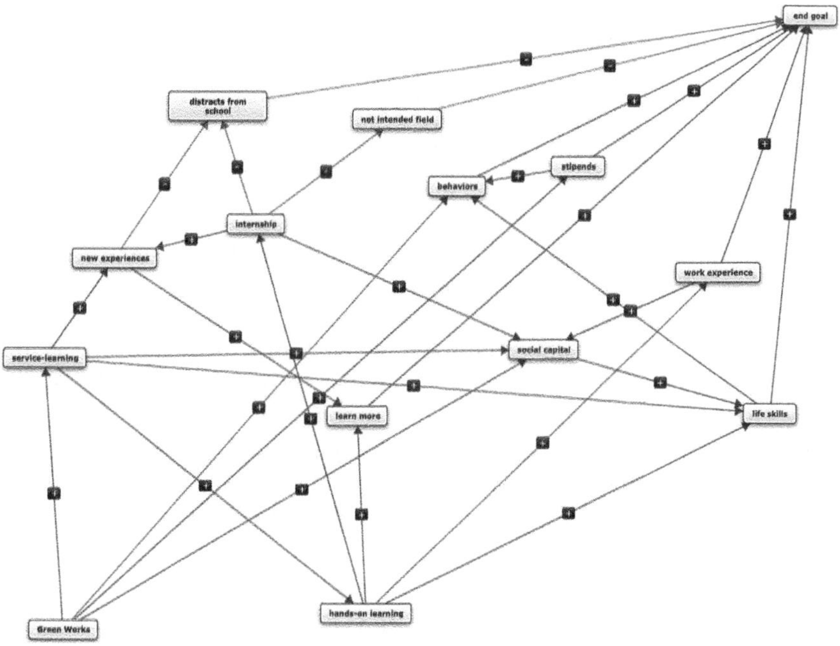


**Figure 3.1** Student cognitive map.

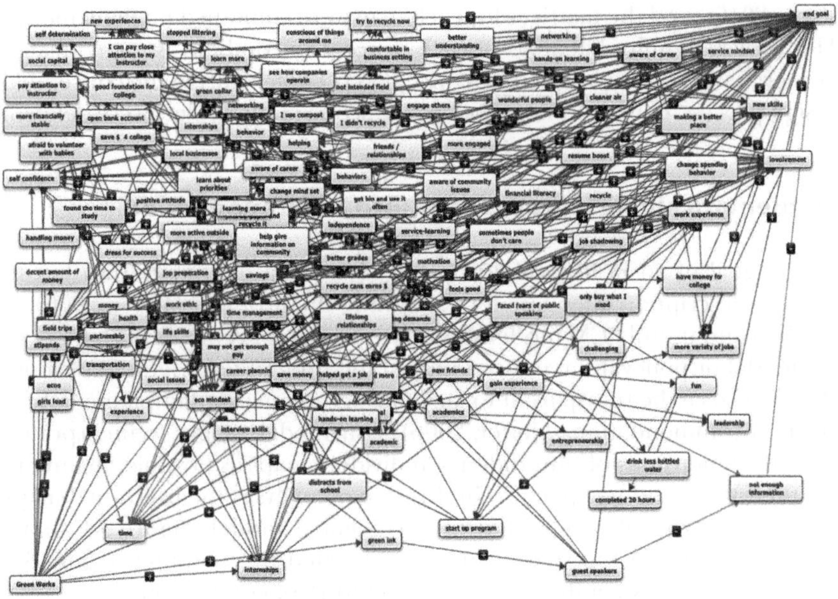

**Figure 3.2** Combined cognitive maps.

## Test, Interpretation, and Validation of Model Results

To ensure the model worked correctly, several tests were run using FCMapper (Bachhoofer & Wildenberg, 2009). The results were compared with the results of the questionnaire for consistency. The scenario testing procedure followed Jetter (2006). Using the literature and the questionnaire results, the evaluator formulated hypotheses to test. Each hypothesis was a set of concepts activated (or, more specifically, set to equal 1 in the FCMapper software). The FCMapper software multiplied the model matrix by an initial state vector in which selected concepts equal 1 and all others equal 0. The result was a new state vector which was again multiplied by the matrix, over and over again, until the matrix values stabilized. The final state vector of the hypothesis testing was compared with the baseline final state vector. To further test the stability of the model, concepts were activated at positive and negative extremes. The extreme results were compared with the data from the questionnaires to make sure that the model was behaving in expected ways.

## Results

Questionnaire results demonstrated how passionate the students were about Green Works. They were doing better in their high school classes (on a scale of 1 to 5, with 5 = strongly agree); students rated "I am doing better in my academic classes" a 4.25/5.00. Using the same scale, 11 out of 14 students either agreed or strongly agreed that Green Works had helped them change their behavior, and 14 out of 14 students agreed or strongly agreed that they thought about the future differently after participating in Green Works. When asked about their preparation for employment, 14 out of 14 either agreed or strongly agreed that they are now better prepared for employment than before Green Works. Table 3.1 summarizes the results of the questionnaire.

The strength of FCM lies in running scenarios to model how different program elements interact. The matrix resulting from the combined maps was loaded in FCMapper (Bachhoofer & Wildenberg, 2009), with several types of analysis run. The following list shows the findings of the quantitative analysis and provides brief interpretation.

- **Number of concepts** = 42
- **Number of connections** = 307
- **Number of self-loops** = 5 (concepts connecting to themselves)
- **Number and percentage of transmitter** ($n = 2$), receiver ($n = 1$), and ordinary concepts ($n = 39$): transmitter concepts lack in-degree; receiver concepts lack out-degree. The low number of transmitter concepts meant students felt self-determination when it came to the ways in which Green Works contributed to their end goal. Transmitters are concepts over which students did not sense having control; in this case, work experience and volunteer experience, that involved negotiation with other parties.

**TABLE 3.1 Survey Responses**

| Measure | *N* Valid | Mean | Mode | Std. Deviation |
|---|---|---|---|---|
| I am doing better in my academic classes. | 12 | 4.250 | 5.00 | .86603 |
| I have changed my behavior. | 14 | 4.00 | 4.00 | .67937 |
| I think differently about my future. | 14 | 4.57 | 5.00 | .51355 |
| I feel better prepared for employment. | 14 | 4.50 | 4.00 | .51887 |
| I have met and worked with a number of different adults from many different fields. | 14 | 4.36 | 5.00 | .84190 |
| I am handling my money differently. | 14 | 4.21 | 5.00 | .80178 |
| I am setting goals for myself and working towards them. | 14 | 4.36 | 4.00 | .63332 |
| I have gotten involved in more neighborhood and community activities. | 14 | 3.8571 | 4.00 | 1.6732 |
| 5-point scale: 1 = strongly disagree; 2 = disagree; 3 = neutral; 4 = agree; 5 = strongly agree | | | | |

- **FCM density index**. The number of connections divided by number of concepts to the power of 2. The lower the number, the lower the density: $307/42^2$ or 0.174.
- **FCM complexity index**. The number of receiver concepts $R$ divided by the number of transmitter concepts $T = 1/2$, or 0.5.
- **Concept in-degree.** The total amount of effects received by a concept irrespective of the strength of the effect. This revealed the students' perceptions of the personal effect of Green Works programs. The concepts with the highest in-degree were the end goal set by the students (20.75); hands-on learning ($n = 8.25$); savings ($n = 6.50$); motivation ($n = 6.25$); career awareness ($n = 5.75$); building a résumé ($n = 5.50$); service mindset ($n = 5.25$); ecological mindset ($n = 5.25$); learning more ($n = 5.25$); and increased sense of engagement with the environment and their communities ($n = 5.25$).
- **Concept out-degree** is the combined strength of effects a concept has on any number of other concepts. The out-degree concepts lead to the in-degree concepts. The connections from out-degree to in-degree to out-degree eventually led to the students' stated end goal. The concepts with the largest out-degree were service learning ($n = 20.75$); Green Works programming ($n = 12.75$); learning about leadership ($n = 12.50$); career planning ($n = 9.75$); internships (which were strongly associated with service-learning) ($n = 7.50$); social issues ($n = 7.25$); and being involved with the environment and their local communities ($n = 7.25$).
- **Centrality** combines the values of in-degree and out-degree. The highest concepts by centrality were service learning ($n = 22.50$); the students' end goals ($n = 20.75$); leadership ($n = 16.25$); and Green Works programming ($n = 14.75$).

The next step was to use FCMapper, which runs squashing functions to test scenarios. The first scenario tested was to maximize the top out-degree concepts to see what changes that would yield. Adding maximum positive change to Green Works programming, service learning, leadership, career planning, internships, and community engagement, stabilized in 11 iterations and led to multiple maximum positive changes for the students as summarized in Figure 3.3.

A second scenario we ran that increased all Green Works programming (formal programs plus learning strategies such as service learning and internships) stabilized in 13 iterations and yielded maximum positive change in the following areas: self-determination, leadership, networking, gaining information, self-confidence, and understanding social issues (see Figure 3.4). It had a medium positive effect on science academics, career

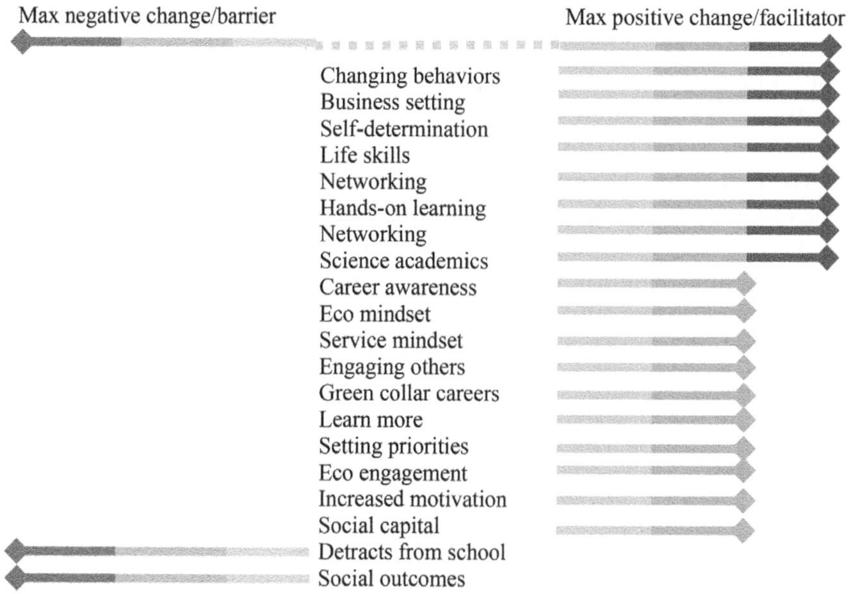


**Figure 3.3** Scenario: Maximum positive change to top out-degree concepts.

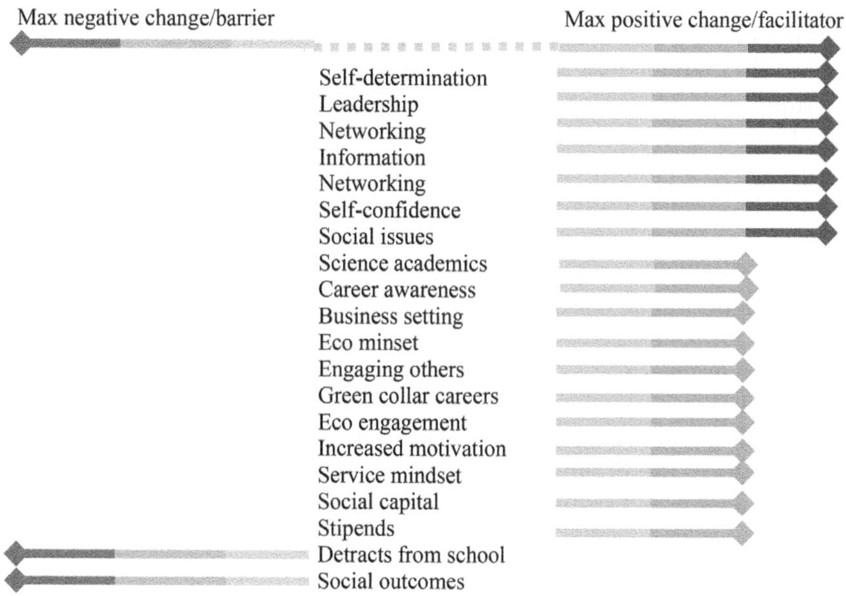


**Figure 3.4** Scenario: Maximum positive change to Green Works programs.

awareness, changing behaviors, feeling comfortable in a business setting, eco mindset, engaging others, awareness of green collar jobs, being involved in the environment and communities, learning about priorities, increasing environment-related learning, social capital, and service mindset.

To understand the impact of service learning on the model, a scenario was run activating the service learning concept at -1. Stopping service learning in Green Works would have strong negative effects on the program in the following areas: academics, career awareness, positive behaviors, career planning, comfort in the business setting, self-confidence, internships, involvement, leadership, learning about priorities, increased learning, life skills, level of engagement, networking, service mindset, and understanding social issues.

## Discussion

The FCM evaluation focused on the following two questions: How does Green Works help urban teen participants reach their goals? and How integral is service learning to the overall Green Works program experience?

## HOW GREEN WORKS HELPS STUDENTS REACH GOALS

To understand how Green Works helps student participants reach their goals, it is useful to look at the top student-identified concepts most strongly supporting their end goal. The end goal most often was high school graduation and transition to employment or post-secondary education. A streamlined conceptual model of Green Works showed conceptual and causal linkages between programming, student-identified concepts, and the student-identified end goal (see Figure 3.5). This model supports the anecdotal evidence of Green Works' success. In the section below, each of the concepts will be discussed.

### *Changing Behaviors*

Green Works upholds a commitment to serving only what we identify as quality, highly motivated participants rather than having large numbers of students in the program. Based on the above analysis of the individual maps, students had strong connections between changing their behaviors and meeting their self-identified end goal. Most of the changed behaviors related to a combination of environmental stewardship, talking with friends and family about changing their behaviors toward the environment, and saving money.

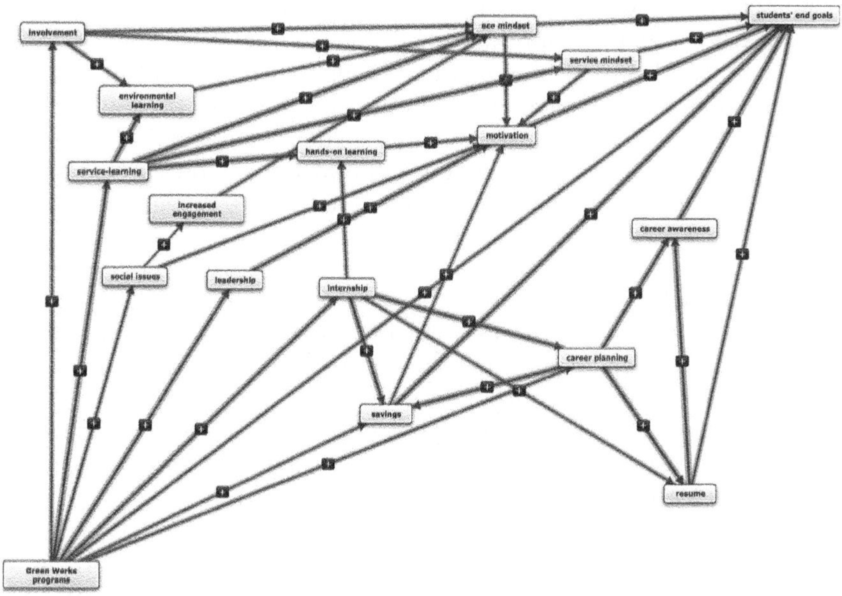


**Figure 3.5** Green Works simplified FCM showing concepts with highest degree of centrality.

***Business Setting***

This concept, in its expanded form, is feeling comfortable in a business environment. For many participants, Green Works provided first experiences and unexpected experiences. Many of the students in the program had only limited experience with a business setting, therefore this concept was important to their sense of career awareness and meeting their goals.

***Self-Determination***

Concepts which were combined into this category included setting and reaching goals, working to make things better, maturity, prioritizing, being persistent, exploring interests, and gaining confidence in my goals. The students connected concepts related to self-determination to their self-identified goal.

***Life Skills***

The students connected life skills directly with reaching their goals. The types of concepts, leading to life skills were earning stipends, opening checking accounts, saving money, and learning about financial management. Most young adults in the United States do not receive adequate personal finance training. While unfortunate, students from middle-income families can usually overcome this setback. Their families have the resources to serve as a safety net for the young adults to move back home, pay off their debt, and begin to save; or their families have the funds to directly subsidize their expenses. Green Works students, however, do not have the luxury of this safety net. Learning to manage their money and learning the reality of how expensive it is to live on your own and the true cost of owning a car is vital. Because the students earned stipends, they actually had money to manage.

#### *Science Academics*

Most students included some form of increase in science learning or literacy on their map. Many of the students referred to this concept as academics, so the combined concept became science academics. Green Works students increased their science literacy and had many opportunities for the types of learning that work (e.g., hands-on learning, service learning, and applied learning).

#### *Service Mindset and Eco Mindset*

These concepts were fairly common across student maps and were a bit mysterious. They were typically connected to self-determination concepts and the end goal statement. The concepts stemmed directly from the service learning projects and the environmental stewardship students connected with through Green Works. The students often remarked that they had not worked on projects in their neighborhood before Green Works. Most had never planted a plant or participated in community information fairs. The service learning projects seemed to empower the students, giving them a sense that they could influence their communities and that they could make a difference.

#### *Career Awareness*

The economy has changed radically in the past 30 years. The manufacturing jobs available to high school graduates have moved overseas or been automated. A college degree is not a guarantee of employment either: over 50% of recent college graduates are currently underemployed in jobs that do not require their education (Smith, LaVelle, Lyons, & Silverstone, 2016). Income is the best predictor of college completion (Thomas, 2013) and yet between 2001 and 2011 poverty increased 40% in the Midwest (Thomas, 2013). For these reasons and more, a traditional 4-year college degree often is not the best path for Green Works students. There is a bright spot in the growth of midskill jobs (i.e., jobs requiring a 2-year associates degree, occupational license, or certification). These jobs are in demand, and current training is not meeting that demand. Numerous 2-year degree holders in many high-demand occupations now earn more than college graduates (Goodwin, 2012). Those with a 2-year degree in STEM fields have greater average lifetime earnings than college graduates in most other careers (Goodwin, 2012).

#### *Increased Sense of Engagement With the Environment and Their Communities*

In their maps, students referred to involvement, engagement, or making a difference. Green Works has a goal of involving youth in the world outside their immediate neighborhood. The Year 1 program incorporates eight field trips. Students took a boat ride on the Missouri River, toured the wastewater and water treatment plants, visited a landfill, visited recycling facilities, and toured LEED (Leadership in Energy and Environmental Design)-certified buildings. Through service-learning projects, students had a number of opportunities to be involved in clean-ups, planting projects, and community forums. Many of these opportunities were also located outside the normal boundaries of the students' day-to-day lives. Finally, students participated in internships at organizations spread over the community, although within access of the public bus system.

#### *Internships*

Not all of the students had participated in an internship yet, so internships were not incorporated in every map. For the students completing their second year, the summer internship was an important factor in realizing their goals. Green Works promotes internships for all of its students and works hard to secure summer internships for Year 2 students. In 2013, the summer employment rate for 16–19 year old African Americans was just 19% (Hall, 2013), yet research indicates significant outcomes for young adults who work during the summer, including increased high school graduation rates, increases in learning, a smoother transition to the labor market, and higher weekly earnings for up to 15 years after high school graduation (Orrell & Ouellette, 2008). Green Works internships provide meaningful work and a supportive first job for students. The students connected internships with self-confidence, self-determination, their end goal, social capital, and career awareness.

## Implications for Other STEM Service Learning Projects

According to the students' maps, service learning in an informal learning setting like Green Works made a difference for students. Service learning best practices include involving STEM professionals as mentors, advocates, and connectors for the students; locating a majority of the service learning projects in the students' communities; choosing significant projects that tie directly to the curriculum and city priorities; and finding ways for the students to write and reflect on the project and share the results with their friends and family.

### *Informal Science Learning*

Green Works engages students in informal learning programs for the entire school year and then throughout the summer. Because the students tend to move frequently, have inconsistent cell phone service, and no or limited Internet service, full-year involvement fosters engagement and connection. Green Works also partners with popular trusted teachers and community members to help with recruiting students. They recruit from a 150-block target area, and within that area they partner with churches, schools, and community and neighborhood organizations to help sell the program to potential students. Students who begin participating in Green Works see science learning increase due to the positive learning environment, which includes small total enrollment and a low student-to-teacher ratio. Green Works accepts 20 new students into the Year 1 program and then continues to work with 20–25 students in Year 2 and beyond. Two teachers work together in each Green Works program. The Green Works teaching staff holds degrees in either environmental science/studies or education with extensive experience in the classroom. In addition, all staff members and many volunteers are environmental activists advocating for increased public transportation, limiting bottled water, fighting invasive plant species, and improving recycling rates.

### *Clearly Defined Outcomes*

All of the Green Works programs have clearly defined and achievable outcomes for the student participants. During the program, our staff reviews the outcomes periodically with the students, which helps foster self-determination and increases the students' engagement with the program.

### *Strategies for Career Exploration Specific to Urban Teens*

Midskill jobs offer an excellent career path for Green Works students. The students, however, need to know about these jobs. They are not represented in their urban neighborhoods nor in the media, and more times than not, the high school counseling services provided to students are not adequate. For Green Works students, the career exploration in which they participate can have a profound impact on their career choice, their long-term earning potential, and consequently, their ability to support a family and buy a home. It is of little surprise that career awareness was predominant on students' maps and ranked high as an outcome in the FCM scenarios.

Green Works facilitates career exploration by working closely with 52 environmental or STEM organizations. These groups provide guest speakers, field trip hosts and sites, facilities, and curriculum materials. The students meet, interact with, and work alongside a large number of individuals who work on behalf of the environment. Green Works builds these partnerships so that students have access to knowledge of a variety of different careers available. Many of these adults become mentors and advocates for our students.

***Importance of Stipends***

Students earn them through reaching goals. As the students progress through Green Works programming, the stipend remains important to them, but they begin to understand that what they are learning, the connections they are making, and the experiences that they gain are important as well. Stipends are incredibly important to program success for two reasons. Firstly, stipends are the main reason a majority of students initially sign up for Green Works. Secondly, when students' lives get hectic and they need to drop out of activities, they usually have support from their families to stay involved with Green Works because their involvement is viewed as a job. Students and their families do not always understand the importance of social capital connections, financial management, and workplace-skill development for the students. What they do understand, however, is that the student is earning money. Often this additional income helps pay family expenses.

## CONCLUSION

Green Works is making a lasting difference in the lives of the students who participate in the programming. Through Green Works, the students are increasing their science literacy, changing their behaviors toward the environment, gaining life skills, increasing their social capital, and spending hours engaged in meaningful service learning and internships. Informal learning opportunities not associated with formal education, that is, long-term with committed professional staff, are not available for every inner-city teen. Informal learning experiences are also not often evaluated or written about because their funding can be precarious, and the main focus is on keeping the doors open and supporting the teens. It is also no secret that the United States needs more adults who complete postsecondary education and build careers in science, technology, engineering, and math (STEM) fields. All Green Works students graduate from high school, have positive early career experience in STEM, and many go on to postsecondary education.

## ACKNOWLEDGMENTS

Green Works in Kansas City would like to acknowledge funders from the past seven years whose contributions have made the program possible: M.R.

and Evelyn Hudson Foundation, Metropolitan Community College, State Farm Insurance, Together Green/National Audubon Society, Environmental Protection Agency, Kansas City Power and Light, and Harley Davidson Foundation.

## REFERENCES

Arnett, S. E., Kitchel, A., & Cannon, J. G. (2011). Experiential learning curricular development model for stimulating student interest in green collar careers. *Online Journal for Workforce Education and Development, 5*(1), 3.

Bachhoofer, M. & Wildenberg, M. (2009). Welcome to FCMappers! *FCMapper.* Retrieved from http://www.fcmappers.net

Billig, S. H. (2010). Why is service-learning such a good idea? Explanations from the research. *Colleagues Magazine, 1*(2), 4–7.

Bridgeland, J. M., DiIulio, J. J., Jr., & Morison, K. B. (2006). The silent epidemic: Perspectives of high school dropouts. *ERIC.* Retrieved from http://eric.ed.gov/?id=ED513444

Bridgeland, J. M., DiIulio, J. J., Jr., & Wulsin, S. C. (2008, April). Engaged for success: Service-learning as a tool for high school dropout prevention. Retrieved from http://files.eric.ed.gov/fulltext/ED503357.pdf

Campbell, C. (2009). Evaluation report: The environmental connection opportunities for students (ECOS) program of Green Works in Kansas City, Inc. *Green Works in Kansas City.* Retrieved from http://www.greenworkskc.org/changing-their-future/

Cole, J. R., & Persichitte, K. A. (2000). Fuzzy cognitive mapping: Applications in education. *International Journal of Intelligent Systems, 15*(1), 1–25.

Cooney, K. (2010). The promise and pitfalls of employer-linked job training for disadvantaged workers. *Administration in Social Work 34,* 27–48.

Corporation for National and Community Service. (2006). *Educating for active citizenship: Service-learning, school-based service, and civic engagement.* Washington, DC: Corporation for National and Community Service. Retrieved from http://www.nationalservice.gov/sites/default/files/documents/SL_briefing.pdf

Corwin, K. (2014). Annual report: September 2013–August 2014. *Green Works.* Retrieved from http://www.greenworkskc.org/wp-content/uploads/2014/11/GW-annual-report-Year-7.pdf

Corwin, K. (2015a). Meet some of our students. *Green Works in Kansas City.* Retrieved from http://www.greenworkskc.org/changing-their-future/meet-some-of-our-students/

Corwin, K. (2015b). About us. *Green Works in Kansas City.* Retrieved from http://www.greenworkskc.org/about-us/

Corwin, K. (2015c). Changing their future—Student and community outcomes. *Green Works in Kansas City.* Retrieved from http://www.greenworkskc.org/changing-their-future/

Corwin, K. (2015d). Programs. *Green Works in Kansas City.* Retrieved from http://www.greenworkskc.org/programs/

Creswell, J. W. (2007). *Qualitative inquiry and research design: Choosing among five approaches.* Thousand Oaks, CA: Sage.

Falxa-Raymond, N., Svendsen, E., & Campbell, L. K. (2013). From job training to green jobs: A case study of a young adult employment program centered on environmental restoration in New York City, USA. *Urban Forestry & Urban Greening, 12*(3), 287–295. doi:10.1016/j.ufug.2013.04.003

Goodwin, B. (2012). Research says don't overlook middle-skill jobs. *College, Careers, Citizenship, 69*(7), 86–87.

Gray, S. (2016). *Mental Modeler* (web-based application). Retrieved from http://www.mentalmodeler.com/#download

Hall, K. G. (2013, August 29). Teen employment hits record lows, suggesting lost generation. *McClatchy DC.* Retrieved from http:// http://www.mcclatchydc.com/news/nation-world/national/economy/article24755047.html

Hamilton, G. (2002). *Moving people from welfare to work: Lessons from the national evaluation of welfare-to-work strategies.* Washington, DC: U.S. Department of Health and Human Services and U.S. Department of Education.

Jetter, A. J. (2006). Fuzzy cognitive maps for engineering and technology management: What works in practice. *PICMET 2006 Proceedings,* Istanbul, Turkey. doi:10.1109/PICMET.2006.296648

Jetter, A., & Schweinfort, W. (2010). Building scenarios with fuzzy cognitive maps: An exploratory study of solar energy. *Futures, 43,* 52–66. doi:10.1016/j.futures.2010.05.002

Kielsmeier, J. C., Root, S., Pernu, C. & Lyngstad, B. (2010). Growing to greatness 2010: The state of service-learning: Introduction: Reclaiming the learning in service-learning. *NYLC.* St. Paul, MN: National Youth Leadership Council.

Konopnicki, P. (2009, November/December). Sustainability: The next 21st century workplace skill. *Techniques,* 44–47. Retrieved from http://www.vbschools.com/TCE/pdfs/Sustainability%20Article%20Nov-Dec_09_feature_PAGES.pdfNov_Dec_09_feature_SPREADS.pdf

Kosko, B. (1986). Fuzzy cognitive maps. *International Journal of Man–Machine Studies, 24*(1), 65–75.

Kouri, R., & Clarke, A. (2012, May 6). Framing 'green jobs' discourse: Analysis of popular usage. *Sustainable Development.* doi:10.1002/sd.1526

Missouri Department of Elementary and Secondary Education (MDESE). (2014). *Kansas City public schools profile.* Retrieved from http://mcds.dese.mo.gov/quickfacts/SitePages/DistrictInfo.aspx?ID=__bk8100030043008300030073008300

Pinderhughes, R. (2007). *Green collar jobs: An analysis of the capacity of green businesses to provide high quality jobs for men and women with barriers to employment.* Berkeley, CA: City of Berkeley Office of Energy and Sustainable Development.

Orrell, B., & Ouellette, M. (2008). *Building effective summer youth employment programs.* Leavenworth, KS: ICF International.

Root, S., & Billig, S. (2008). Service-learning as a promising approach to high school civic engagement. In J. Bixby & J. Pace (Eds.), *Educating democratic citizens in troubled times: Qualitative studies of current efforts* (pp. 107–130). Albany: State University of New York Press.

Smith, D., LaVelle, K., Lyons, M., & Silverstone, Y. (2016). *The gig experience: Unleashing the potential of your talent and your business.* Retrieved from https://

www.accenture.com/t20160512T073844__w__/us-en/_acnmedia/PDF-18/Accenture-Strategy-2016-Grad-Research-Gig-Experience-Unleash-Talent.pdf#zoom=50

Thomas, O. (2013, December 10). A poverty, not education, crisis in U.S. *USA Today.* Retrieved from http://www.usatoday.com/story/opinion/2013/12/10/education-poverty-international-student-assessment-column/3964529/

U.S. Census Bureau. (2012). *Quick facts: Kansas City, Missouri.* Retrieved from http://www.census.gov/quickfacts/table/PST045215/2938000

White, S., & Walsh, J. (2008). *Greener pathways: Jobs and workforce development in the clean energy economy.* Retrieved May 4, 2009, from http://apolloalliance.org/wpcontent/uploads/2009/01/greenerpathways1.pdf

Yamauchi, L. A., Billig, S. H., Meyer, S., & Hofschire, L. (2006). Student outcomes associated with service-learning in a culturally relevant high school program. *Journal of Prevention & Intervention in the Community, 32*(1/2), 149–164. doi: 10.1300/J005v32n01_10

CHAPTER 4

# SERVICE LEARNING EXPERIENCE EFFECTS ON HIGH SCHOOL STUDENTS' LEARNING OF SCIENCE PRACTICES

**Angela Chapman**
***University of Texas–Rio Grande Valley***

**Allan Feldman**
***University of South Florida***

## ABSTRACT

Inequities in U.S. schools have resulted in a disproportionate number of girls, students of color, and those who live in poverty being marginalized in science. Several factors may lead to marginalization, including gender, race, ethnicity, socioeconomic status, speaking English as a second language, and the need for special education services. These factors have been attributed to the science achievement gap existing between girls and boys, as well as students of color and white students. This gap has led to the underrepresentation of these students in careers related to science, technology, engineering,

*Science and Service Learning,* pages 73–97

and mathematics. The purpose of this study was to determine to what extent marginalized students were affected by participation in an intervention combining authentic science and school-based service learning at an urban high school. Marginalized students engaged in authentic science by performing original scientific research in collaboration with faculty from a research-intensive university. Using a mixed-method research design, we offer an intervention strategy that combines authentic science experiences and service learning as part of regular science classroom instruction as a means of engaging marginalized students in the learning of science.

Inequities in U.S. schooling have resulted in a large number of girls, students of color, and those who live in poverty to be marginalized in science. Many of these students are unofficially tracked into science classes that do not focus on higher-level cognitive processes that promote critical thinking or improve academic achievement (Oakes, 2000). In addition, students in these classes often have limited opportunities to engage in high-quality science instruction (Barton, 2001; Darling-Hammond, 1999; Oakes, 1990). As a result, these students may become marginalized since they are often disengaged, lacking in self-efficacy about their science abilities, and/or view science as irrelevant and boring (Barton & Yang, 2000; Basu & Barton, 2007; Seiler, 2001; Zacharia & Barton, 2004). While marginalization is a multifarious concept, several factors have been identified (Buxton & Lee, 2010).

This chapter reports on the effects of an intervention that combines authentic science and service learning in the context of an algal biofuel research project (ABFP). Students at an urban high school, all of whom possessed at least one marginalizing factor, engaged in authentic science by performing original, scientific research in collaboration with faculty from a research-intensive university. Specifically, students were investigating optimal algal growth conditions for use as a biofuel. The purpose of this study was to determine how and why marginalized students were affected by participation in this project. The research questions guiding this inquiry were

1. How did involvement of students in an intervention that integrated authentic science and school-based service learning affect their perceptions?
    a. Did students perceive the experience to be one of authentic science?
    b. How did participation affect students' perceptions about their research skills and engagement as a scientist?
2. How did involvement of students in an intervention that integrated authentic science and school-based service learning affect their learning of science content?

## SERVICE LEARNING THROUGH AUTHENTIC SCIENCE EXPERIENCES

The most recent release of the Next Generation Science Standards (NGSS) calls for the implementation of eight scientific practices in K–12 education (NGSS, 2013). These practices allow students to engage in scientific investigation as a way of learning the content and the requisite skills. The eight scientific practices offered by the NRC's Science Framework for K–12 Science Education are (a) asking questions; (b) developing and using models; (c) planning and carrying out investigation; (d) analyzing and interpreting data; (e) using mathematics and computational thinking; (f) constructing explanations; (g) engaging in argument from evidence; and (h) obtaining, evaluating, and communicating information.

The science practices described by the NGSS are the knowledge and skills needed to engage in science in ways similar to that of scientists. When K–12 students engage in science activities that closely parallel the practices of scientists, they are said to participate in authentic science experiences (Chinn & Malhotra, 2002; Hsu, Roth, & Mazumder, 2009). Chinn and Malhotra (2002) state that this includes generating research questions, designing studies, making observations, explaining results, developing theories, and studying research reports. These views of authentic science reflect the NGSS' description of science practices.

While multiple models of authentic science have been offered, Buxton (2006) categorizes authentic science experiences as canonical, youth-centered, or contextual. Canonical authentic science experiences focus on the authenticity of the science experience itself (Chinn & Malhotra, 2002). Their definition requires having students "modeling what scientists do, providing problem-solving activities that are accessible to students and motivate them to explore the natural world" (Buxton, 2006, p. 699). In contrast, youth-centered authentic science experiences focus on the students' voice (Barton, 1998; Eisenhart, 2001). For example, Barton's (1998) study of homeless children is framed from the position that science is "a process which can be constructed out of any pursuit, experience, or interest that the participants bring" (p. 383). Lastly, Buxton (2006) suggests that one model of authentic science can be viewed contextually as a hybrid between the canonical and youth-centered models as it focuses both on the authenticity of science instruction as well as the student perspective. He argues that in a contextual view of authentic science, the values and beliefs of students as stakeholders should be recognized and students should be allowed a voice in what they investigate. Also, this view typically includes the social issues surrounding the content. As a result, this view is a blend of canonical authentic science that has been described as "scientist's science" reflective of more traditional Western science and student-centered authentic science that puts greater emphasis on student interests. Buxton argues that the contextual view can empower marginalized students. We propose that authentic science activities modeled after this hybrid perspective can provide students with the opportunity to simultaneously engage in high-quality science learning and service learning.

While multiple definitions of service learning are present in the literature, we consider service learning here as a form of instruction that connects classroom learning objectives with a community or social need; students are allowed to develop a sense of civic responsibility and more empathetic view of their community. The benefits of service learning to students are that it (a) positively influences academic performance, (b) improves motivation to learn science by making science content more personally relevant, (c) promotes student ownership of learning, and (d) helps them to develop a sense of civic responsibility in that they can make a difference in their community (Billig, 2000; Simmons, 2000).

Whereas service learning has been argued as a form of high-quality instruction for K–12 and postsecondary students (Billig, 2000), less is known about the impact of science service learning for high school students. However, science service learning for college students has been reported (Barton, 2000; Kamela & Gammon, 2011; Simmons, 2000). For example, Kamela and Gammon (2011) demonstrated how students at a university used a mobile science lab to engage young children in an informal science setting. This study describes how to engage nonscience majors in the learning of science through student developed, service learning projects. Simmons (2000) describes ways to engage college students in service learning in the context of environmental science issues. We report how college students shifted their beliefs about the value of what they were learning because they recognized it as relevant and having a practical application.

We see the intervention studied, the algal biofuel project (ABFP), as a different way to think about and enact service learning. Often, service learning is conceived as activities occurring outside of the formal educational setting, hence the use of the label "community service learning." In our study, high school students engaged in service learning as part of regular classroom instruction. We therefore refer to it as school-based service learning. In the ABFP, learning occurred within the school and connected the students to their global and local communities. Although environmental problems, such as the need for alternative energy sources, the demand for clean water, or the production of biofuels from wastewater, which are the foci of the ABFP, are global in scope, they have a constant presence in the daily lives of high school students. Environmental science as school-based service learning when implemented as an authentic research-based project can empower students by helping them develop a sense of how they can make a difference in their communities.

## METHOD

The study took place at an urban high school in the southeastern United States serving a diverse student population. At the time of the study, the school had approximately 1,600 students of whom 71% were Black, 14% were White, 11% were Hispanic, and 2% were Asian. In addition, 64% were eligible for a free or reduced-fee lunch; 12 students in an honors Marine Science class participated in this study. The class was open to enrollment by neighborhood students as well as those attending the school for any of its magnet programs. Historically, the school, while not in its current building, was one of two all-black schools in the city prior to desegregation. It continues to serve the surrounding historically African American neighborhood. Table 4.1 summarizes the demographics of students participating in the study. The high school teacher provided us with information about the students' gender, magnet status, and socioeconomic status, while the students self-reported their ethnicity. We used pseudonyms throughout the study for all participants. The University's IRB and the school district's office of research approved our study. All participants and their parents or guardians provided consent to be in the study.

In order to create a situation in which the students experienced authentic science activities, we helped set up a partnership between the teacher, her students, and the Algal Biofuels Research Group (ABFRG)[1] from a local research-intensive university, which we refer to as the University. Although the ABFRG was located in a department of environmental engineering, their research resembles what the *Framework for K–12 Science Education* (NRC, 2012) refers to as science rather than engineering practices. We therefore refer to their research as science, not engineering. Over a 2-month period the high school students engaged in science activities similar to those practiced by members of the ABFRG by investigating optimal growth conditions for the production of biofuels in synthetic wastewater. Their teacher, Mrs. Bodin, served as a facilitator and ensured students collected data and

**TABLE 4.1 Demographics of Students Participating in ABFP**

| Demographic | Category | # Students |
|---|---|---|
| Gender | Male | 8 |
| | Female | 4 |
| Ethnicity | White | 2 |
| | African American | 3 |
| | Hispanic | 5 |
| | Multiracial | 1 |
| SES | High | 4 |
| | Low | 8 |

maintained the bioreactors. She had never participated in this type of project, so she was a novice, along with her students.

As described previously, we developed this intervention using the contextual model of authentic science (Buxton, 2006). The Multiple Outcome Interdisciplinary Research and Learning (MOIRL) model (Feldman, Chapman, Vernaza-Hernández, Ozalp, & Alshehri, 2012) guided our design of the intervention. The MOIRL model allows for learning to occur as a result of different stakeholders engaging in scientific and educational activities through their legitimate peripheral participation (Lave & Wenger, 1991) in research communities of practice (Feldman, Divoll, & Rogan-Klyve, 2009, 2013; Wenger, 1998). While the stakeholders included the two authors of this article, the ABFRG from the University, the teacher, Mrs. Bodin, and her students, our primary focus in this chapter is on the learning of the high school students during a service learning project taking place as part of regular classroom instruction in an Honors Marine Science course.

The primary objective for Chapman (2013) was to use the data for her dissertation. She is also passionate about her goal to reduce inequities found in U.S. schools. The objective of Feldman, who was Chapman's dissertation chair, was to help her complete her doctoral work. He also shares her commitment to promoting social justice. This study is part of his larger research agenda on how people learn to do scientific research (Feldman et al., 2009, 2013). The ABFRG, consisting of two doctoral students, Grace and Katia, and their advisor Professor Berber, had several objectives. Grace is interested in science education and included aspects of the intervention in her dissertation. Professor Berber has a long-term interest in outreach to the K–12 setting. The ABFRG hoped that the high school students' research would help their research by investigating the effects of variables such as light and pH on algae growth. Mrs. Bodin saw this as an opportunity for her high school students to improve their educational experience while learning about the role of photosynthetic algae as part of the marine science class instruction.

The use of the MOIRL model to structure the intervention allowed for multiple ways to enact the project as a school-based service learning experience. As noted above, Professor Berger's long-term interest in outreach to K–12 education provided an opportunity for her undergraduate and graduate students to both do service learning and learn how to incorporate it in their future practice. Likewise, Chapman had her middle and high school students participate in service learning projects as a former K–12 educator. She learned how a project like this one could allow her future preservice teachers to implement authentic science as part of regular classroom instruction. Feldman has a long-standing commitment to environmental education and environmental issues and saw how this type of project could engage preservice and in-service science teachers in school-based service learning. Mrs. Bodin had never engaged her high school students in this type of experience. Thus, she saw this as an opportunity to learn how to engage her current and future high school students in service learning aligned to the course curriculum. Our focus in this chapter is on the outcomes of the high school students. The high school students saw it as an opportunity to make a difference in that they were trying to develop an alternative fuel that would be beneficial to the environment.

Students built bioreactors and cultured algae in a greenhouse on the high school campus. Over a 6-week period, students measured biomass production, pH, temperature, and natural light in an effort to determine optimal growth conditions. Each student group maintained three bioreactors, each with a specific ratio of nitrate and ammonia. During the project, student groups were responsible for maintaining a laboratory notebook of their methods and data as well as orally presenting their findings at the end of the project. Each student maintained a journal that included descriptions of what they were doing in the project as well as writing a research summary. Journals allowed students to reflect on their experience during the project.

There were two phases to the students' research. In the first phase, they were supplied with the research question to determine the best ratio of nitrate and ammonia for optimal algal growth. Phase 1 also provided the students the opportunity to practice skills needed to do the research. In order to determine the best ratio, students needed to compare the findings of all groups. This was accomplished by having students create a growth curve graph showing algal growth under different conditions by plotting algal total solids (*y*-axis) over time (*x*-axis). Mrs. Bodin instructed each group to add their growth curve data to an Excel spreadsheet so that students could generate a growth curve (Figure 4.1) to determine which set of conditions generated the most algae. Students concluded that a 50:50 of nitrates:ammonia provided optimal algal growth. Next, each student group worked collaboratively with Professor Berber and her graduate students to use findings from Phase 1 to generate his or her own research question for Phase 2. There were four groups of students in the class. After discussing the Figure 4.1 findings with Professor Berber, the high school students decided all experiments would be conducted under the 50:50 nitrate:ammonia conditions. One group of students offered to replicate these conditions for Phase 2, which they agreed would serve as the control group. The research questions and hypotheses generated by three groups were

1. How does an extra diffuser affect the growth of algae? If I have an extra diffuser in each of my bottles with the algae and wastewater then the quantity of Chlorella Sp. algae grown will be superior to the rest of the experiments because the Chlorella Sp. algae will grow better with more circulation of the algae and help with gas exchange.

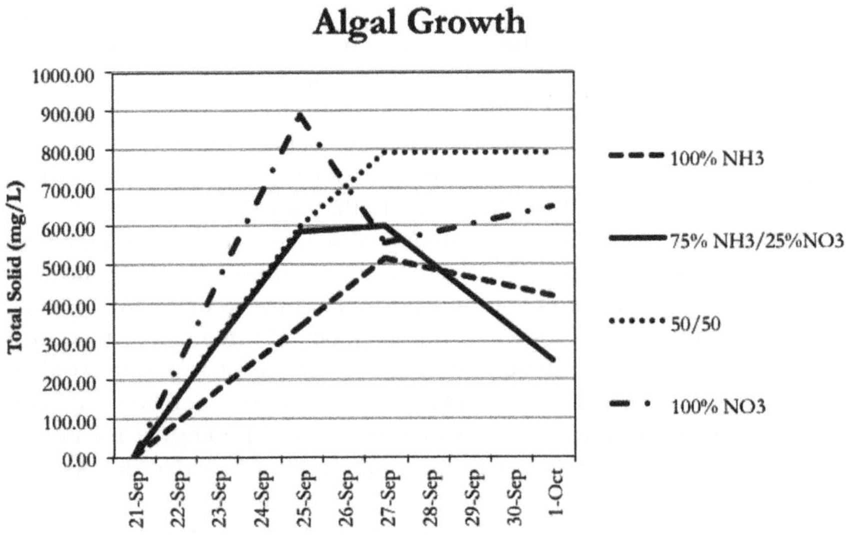


**Figure 4.1** Student generated graph showing algal growth under 100% $NH_3$ conditions.

2. Does the constant light benefit or reduce the growth of algae? If we keep constant light on the algae then the algae will have a steady productive grow.
3. How does increasing and stabilizing the pH affect the growth of algae (Chlorella sp.)? [no hypothesis stated]

The designs of Phases 1 and 2 are important for two reasons. First, students were actively engaged in science as a collaborative endeavor as each group compared its data to the control group data in order to make an evidence-based conclusion regarding their investigation. Second, this process provided students with a voice in what they were investigating while ensuring what they were investigating remained within the parameters of Professor Berber's research objectives. The second phase of the intervention design was developed to reflect a contextually based authentic science model (Buxton, 2006; Feldman et al., 2012).

Another aspect of the intervention involved a field trip to the University near the end of the project. After the students had conducted algal biofuel experiments, but before they had completed the data analysis and presentation of findings, students attended an engineering graduate student symposium, listened to a keynote address from an engineering faculty member, and were given a tour of Professor Berber's lab. During this field trip, students interacted with graduate students and faculty who were diverse in terms of age, gender, and ethnicity.

At the end of the project, the students presented their findings to an audience that included Professor Berber, Mrs. Bodin, school and district administrators, Chapman, and Feldman. This was modeled after typical conference presentations in which a question and answer session followed each presentation.

Our design of the intervention intended to provide the students with the opportunity to make explicit connections between the algae biofuel project, authentic science, and specific scientific practices by researching an important environmental issue. This project provided the opportunity for students to understand that science is a collaborative endeavor. Students were able to make explicit connections between what they were doing and the "real" science of the ABFRG. As previously described, the design of Phase 1 and Phase 2 required the students to collaborate with each other as each group generated data that had to be analyzed collectively by the entire class in order to form an evidence-based scientific conclusion. Students also recognized themselves as collaborators with the Professor Berber's research group. We explain how students can learn that this type of collaboration could lead to the resolution of societal problems, in this case the development of biofuel as an alternative energy while concurrently making high-quality science instruction a more equitable practice.

## DATA COLLECTION

We used a fully mixed, concurrent, dominant status paradigm (Leech & Onwuegbuzie, 2009). Data were collected from observations, interviews, student journals, video presentations, and a research-skills survey (Kardash, 2000). We describe each of these data sources and methods of analysis below.

In order to understand the effects of participation in the algal biofuel project, we conducted interviews of the students after the intervention. Using a semistructured protocol, questions were designed to collect data on what they learned, their thoughts, and their beliefs about the project. We also included questions about their understanding of what authentic science is, how this experience compared to the school science they experienced before the project, and what they learned from the project with regard to content and service to the community. The interviews, journals, presentations, and classroom observations allowed for the identification of shared experiences among participants.

As an observer (Johnson & Christensen, 2013; Patton, 2002), Chapman was able to provide details about the classroom, learning environment, and culture providing context for the data collected from interviews, surveys, tests, and journals. In this role, Chapman was able to produce a detailed recording of experiences that occurred during the project, including level of student engagement; interactions among researchers, students, and teachers; and student roles as scientific researchers.

At the end of the ABFP, each student group delivered a presentation of its findings to an audience of MOIRL stakeholders. These presentations were video recorded and analyzed to determine the students' level of understanding of the scientific content and practices. Students maintained an individual journal, which allowed them to record their experiences during the ABFP. Specifically, the journals were a place for students to explain their experimental results and thoughts regarding participation in the ABFP.

A modified version of Kardash's (2000) research-skills survey was used to measure students' perception of their scientific skills before and after participation in the ABFP. Data were analyzed both quantitatively and qualitatively. Methods of analysis are described next.

## DATA ANALYSIS

We analyzed qualitative data using a two-phase process to determine patterns or categories of emerging themes (Johnson & Christensen, 2013). During the first phase, we read the interview transcripts, journal entries, and classroom observations in order to generate codes from portions of data. This included *in vivo* codes derived from students' direct statements, emotion coding, and simultaneous coding. We also used hypothesis coding based on predetermined codes related to the objectives of the ABFP (Miles & Huberman, 1994). The qualitative data were triangulated with the quantitative survey data regarding the students' perceptions of research skills and perceptions about who can do science.

Quantitative data from the research skills survey and science practice scores were analyzed using descriptive and inferential statistics. The median, average, and frequency distribution were determined from responses to the pre- and postsurvey as well as Phase 1 and Phase 2 summaries demonstrating learning of science practices. Next, the Wilcoxon signed rank test, a nonparametric paired observation test, was used to compare difference in scores on the pre- and postsurvey as well as difference in science practice scores on the Phase 1 and Phase 2 summaries. Statistically significant results were further analyzed for effect size by measuring Cohen's d (Cohen, 1988).

## RESULTS

### Authenticity of Experience

We used two sets of data, the research skills survey and student interviews, to determine whether the students felt that the ABFP was an authentic science experience to them. Four items in the research skills survey were designed to better our understanding of students' perceptions of the authenticity of the experience. These items were:

1. The Algae Biofuel Project will help me to understand how scientists do research.
2. I will participate in the Algae Biofuel Project like a real scientist.
3. Participating in the Algae Biofuel Project will make me want to do research like a scientist.
4. I will feel like a real scientist when I participate in the Algae Biofuel Project.

With regard to these four items, the median score on the presurvey was 3.89 and the median score on the postsurvey was 4.07 out of 5.0. Based on Wilcoxon signed rank-test analysis, there was not a statistically significant difference between the median pre and post responses. The median score on the presurvey was high, indicating that the students expected their experience to be one of authentic science. It appears the instrument may not provide a sufficient range for respondents who score relatively high initially to demonstrate improvement after the intervention. Thus, the minimal difference may be due to a ceiling effect, leaving small room for growth. To determine if the ceiling effect was influencing the results from the pre- and postsurveys, we reanalyzed the data from participants who responded with a 4 or lower Likert scale response on the presurvey. Other researchers have used this approach to examine whether nonsignificant differences were due to the ceiling effect (Morrell & Carroll, 2003; Roberts, Henson, Tharp, & Moreno, 2001). By removing those reporting a score of 5 on the presurvey, the presurvey mean was a 3.46 and the postsurvey mean was a 3.8. This analysis shows a significant increase ($p < .05$) in their perception of research skills and authenticity of the experience. While the small gain in responses on the postsurvey lends uncertainty to the statistical significance, when we triangulate these results with the qualitative data from interviews and journals, we see strong evidence that students believed their participation in the algal biofuel project to be one of authentic science. The qualitative findings are discussed next.

During the interview at the end of the project, students were asked to describe their experience and compare what they did during the ABFP to what they thought scientists do on a daily basis. One student stated that he felt that what he was doing was "very similar" to what a scientist does; that "they probably just use more, um, higher technology than what we do, but it is pretty much the basics... a little, like we tested the algae growth and we got good results and stuff"; and that "they [ABFRG] might want to take notes from what we did into their classroom." When asked how they felt about the ABFRG using their data, one student stated, "It feels really good to know that we could do something like that. And scientists spend their whole day working on something like this and we could come up with that." This statement was representative of 11 of the 12 students in the project.

The survey and interview data support the finding that the students believed their experience to be one of authentic science. First, there is a statistically significant increase in student perceptions about their research skills when the ceiling effect is taken into consideration. Second, students made reference to what they are doing as having a high degree of familiarity with what they perceive scientists are doing. Also, students felt that an individual who is a scientist was recognizing them as a scientist, and they perceived their results from the experiments as useful to the ABFRG.

We wanted to determine if students had engaged in this type of learning in previous science classes. Thus, students were asked to compare how the ABFP compared to what they typically did in science classrooms. For example, Neal stated,

> This was a new experience because usually when we do labs the teacher already has the data and the research, so basically we have to plug it in. But with our experience we had to find it out for ourselves and add to the experiment as we go.

This statement was representative of comments made by all students involved in the project in that the experience was unlike any previous science learning experience.

## Student Perceptions of Research Skills

Quantitative data from the research skills survey and qualitative data from interviews and journals were analyzed to examine student perceptions about their research skills. Perceptions of research skills were measured at the beginning and end of the project using a modified research skills survey (Kardash, 2000). Students were asked to respond to five statements:

- I can create a research hypothesis.
- I can design a scientific experiment.
- I can make observations and collect data.
- I can figure out what the data means.
- I can explain the results of the research to others.

The median response on the presurvey was 24.5 out of 30.0 and the median response on the postsurvey was 27.0 out of 30.0. Wilcoxon signed rank analysis demonstrates a statistically significant increase ($n = 12$, $z = 2.463$, $p = 0.014$) with a large effect size ($r = 0.71$). Student perceptions of their research skills increased after participation in the ABFP.

Students were asked what they did during the ABFP and if it was similar to that of a scientist. One student, Sophia, responded, "I think the roles are the same; we did the same thing we took the same amount of solution, we dried it weighed it and we did the same thing every day to see if it was growing or dying or the same." Another student, Jessie, stated, "I thought I was terrible at creating a hypothesis, but now I'm, like, oh, I can do this." Isabella was asked about her role(s) in the project, to which she responded, "I did mainly the whole getting the data, I did the light meter, I extracted the algae from the bottle, things like that. I tested the pH. I put the things in the oven." When asked why she did that (put things in the drying oven), she stated, "To measure the algae that was left over, the algae without the water." When Isabella was asked what she learned from the project, she responded, "I basically learned a lot, I didn't know about algae and the different types of algae. I didn't know how to bring an experiment together and put it together and actually let it go and test it for a few days. I learned actual experiments." These statements indicate that students recognize their research skills in the context of the ABFP. All 12 students in the project made at least one explicit reference to their ability to collect and analyze data. The frequencies at which students made reference to each skill are summarized in Table 4.2.

The results of the skills survey and the interviews support the conclusion that students had a positive perception about their research skills as it relates to participation in the ABFP. We describe student learning of science practices next.

**TABLE 4.2 Students' Self-Assessment of Abilities to Engage in Research Skills**

| Research Skill | Percentage of Students Who Agreed or Strongly Agreed |
|---|---|
| I can create a research hypothesis. | 87% |
| I can design a scientific experiment. | 50% |
| I can make observations and collect data. | 100% |
| I can figure out what the data means. | 58% |
| I can explain the results of the research to others. | 50% |

## Learning of Science Practices

Quantitative data from journals, interviews, and presentations were coded using a rubric designed to measure the students' understanding of science practices and content as it relates to the ABFP (see Table 4.3). Students were asked to write a summary with regard to each phase of the research project. The summaries were quantitatively scored using the rubric in Table 4.3. Evidence demonstrating learning of the following specific science practices was determined.

1. Ask a research question or create a hypothesis.
2. Collect data.
3. Analyzed data.
4. Explained results of research.

Learning of science practices 1, 3, and 4 was assessed based on written responses (i.e., Phase 1 and 2 summaries) while science practice 2 was assessed based on classroom observations. This assessment strategy allowed us to generate a score (ranging from 0 to14), demonstrating learning of science practices based on participation in the ABFP for each phase of the project. In addition, we used the same rubric to analyze data from classroom observations, interview and presentation data in order to generate a total science practice score. It is important to note that each student group gave an oral presentation of their findings at the end of the project; students in each group took turns presenting different sections of their project. The data from the oral presentation is incomplete in that the absence of a response does not indicate a student lacked understanding of a specific concept. The Phase 1, Phase 2, and total science practice scores are described in the next section and summarized in Table 4.3.

With the exception of "students collected data," there were four levels of coding. A Level 0 indicated the student did not respond. Students who did not respond were excluded from the statistical analysis. A Level 1 indicated the student responded but there was no evidence of an accurate response. Students' Phase 2 and Phase 2 research summaries were analyzed for their ability to (a) ask a research question or create a hypothesis, (b) collect and analyze data, and (c) explain the results of their research. For the item "student collected data," the maximum possible score was 2, and for the other items the maximum possible score was 3. Explaining the results of the research was divided into two components: connecting evidence to the research question and connecting the evidence to a scientific principle. Each component was assigned a maximum score of 3. We show representative qualitative responses, followed by quantitative analysis of student responses.

**TABLE 4.3 Rubric for Learning of Science Practices**

| Scientific Practice | Level | | | |
|---|---|---|---|---|
| | 0 | 1 | 2 | 3 |
| Student asked a scientific research question OR created a hypothesis. | No response | No evidence | Student offers a naïve response, for example, explains what they were going to do but does not phrase it as a research question or hypothesis. | Student proposes a testable RQ or hypothesis that is specific to the study performed by their research group. |
| Student collected data. | Not observed | No evidence of collecting data (from observations and interviews) | Student was observed collecting data during observations OR referred to doing so during the interview | NA |
| Student analyzed data. | No response | No evidence or inaccurate analysis of data | Student offered an incomplete but accurate analysis of data. | Student offered an accurate and complete analysis of data. |
| Student explained results of research. | No response | Does not use data as evidence to support the research question or provides inappropriate evidence that does not support the research question. | Uses the data appropriately but insufficiently as evidence to support the research question. May include some inappropriate use of data. | Uses the data appropriately as evidence to support the research question. |
| A scientific explanation connects the evidence to the original hypothesis/research question and connects the evidence to scientific principles. | No response | Does not provide reasoning, or only provides reasoning that does not link evidence to the research question. | Provides reasoning that links the RQs and evidence. Includes some scientific principles but is incomplete. | Provides reasoning that links evidence to the RQ. Includes appropriate and sufficient scientific principles. |

One student, Jessie, wrote the following Phase 1 summary. "My results for 75% $NH_3$/25% $NO_3$ were the algae growth slowly declined as well as the pH.... the pH declined when the algae growth declined... the algae degrading due to microbes breaking algae down." The first part of the statement ("My results for 75% $NH_3$/25% $NO_3$") was given a research question score of 2, and a data analysis score of 2 as the statement "where the algae growth slowly declined as well as the pH" is accurate but incomplete. Last, the statement "the pH declined when the algae growth declined... the algae degrading due to microbes breaking algae down" was coded as a 2 for connecting the evidence to their research question/hypothesis and a 2 for connecting the evidence to scientific principles. Both parts of these statements are accurate, but incomplete. Jessie's Phase 2 summary was:

> My groups research question was how does diffusers effect algae growth. Our data explains we had some exponential growth but *declined due to possible environmental factors (rain and cloudy skies) & possible self-shading of algae,* our growth increased *after that period possibly due to more light. Our conclusion is that we did better than the control group with growth because at the end we had a bit more growth.* This is so because in our data near the end of the table the graph for the control compared to ours stated we had a bit more growth. (italics and underline added)

The first sentence is a specific, testable research question and was scored as a Level 3. Several parts of the summary were coded as an accurate but partial analysis of the data. These statements are underlined in the above summary. Statements that were coded as a 3 for both aspects of a scientific explanation are italicized. Jessie also demonstrated an ability to analyze the data and explain the results of her data. During her group's presentation, she stated,

> As you can see from the graph we are 75% ammonia and 25% nitrate as you can see at first we had good growth and then suddenly it declined. *We thought it declined because of the 75% ammonia. Like our fish tanks, for example, if we have too much ammonia our fish will possibly start to die. Just like any other organism, algae and water could possibly die off because of too much ammonia.* Here's our data from Phase 2 and you can see our bottles are kind of brown. You can see we had some good growth and some decline *and our conclusion was we had some possible environmental like there was a lot of rain and possibly the algae wasn't taking in as much light and also this decline could also be from self shading which is when the algae clump together and the algae in the front of the bottle shade the algae in the middle, and the algae in the middle can't take in enough light which causes algae in the middle to slowly die. As you can see it came back up possibly those in terminal factors there was more sunlight and the algae could take the light.* (italics and underline added)

The underlined statements were coded a Level 2 for her ability to analyze data, and the italicized statements were coded as a Level 3 (both aspects). During her presentation, she was asked why algae are used as a biofuel (instead of fossil fuels). Her response was "algae takes $CO_2$ out, that's why we would use them. Yes that's true what you said about the fossil fuels but the algae uses the carbon dioxide as food and makes sugars and oxygen and to grow." This statement shows she is connecting algae growth to the scientific principle of photosynthesis and was coded as a Level 3 for offering reasoning regarding connecting evidence to a scientific principle.

Jessie also made significant statements in her postinterview that demonstrated her ability to analyze data and explain the findings. When asked what would happen to the results of their experiment during her interview, she stated, "I think it's possible they could look at it more because algae tends to... self-shade. Self-shading could cause the other algae to die. So you wouldn't want that. You would want more algae to reproduce." Again, she is showing an ability to connect the evidence from their experiment to a scientific principle of self-shading. Overall, Jessie demonstrates a strong and improved understanding of the content.

All student responses were qualitatively analyzed in the same manner and then analyzed quantitatively using Wilcoxon signed rank analysis (see Table 4.4). This allowed us to look for improvements in students' learning of science practices with regard to their Phase 1 and Phase 2 summaries. Students who did not write either a Phase 1 or Phase 2 summary were excluded from the analysis and are indicated by an asterisk in Table 4.4. The median Phase 1 summary was a 9.0, while the median Phase 2 summary was an 11.0. Based on this analysis, there was a statistically significant difference between Phase 1 and Phase 2 summaries ($n = 9$, $z = 2.533$, $p = 0.011$). In addition, Cohen's effect size ($r = 0.844$) suggests large significance (Cohen, 1988).

## Service Learning

As we argued earlier, we conceive the ABFP as a school-based service learning opportunity for the high school students. In this section, we review our data that supports this view and then consider the project as service learning from the perspective of Professor Berber and her students.

We wanted to understand the students' perceptions with respect to the significance of the research that they were doing in the ABFP. That is, did students see their participation as service to the community? First, as part of a pre- and postassessment, the students were asked, "What benefits are obtained from growing algae?" Four common themes emerged from the data: students recognized algae as photosynthetic organisms, algae as an

**TABLE 4.4 Assessment of Students' Learning of Science Practices**

| Student | Phase 1 Summary | Phase 2 Summary | Total Science Practice Score |
|---|---|---|---|
| Simon | 11 | 2* | 12 |
| Jessie | 9 | 13 | 14 |
| Alex | 7 | 8 | 8 |
| Isabella | 9 | 12 | 14 |
| Sophia | 8 | 9 | 9 |
| James | 9 | 11 | 11 |
| Tyler | 7 | 11 | 12 |
| Jared | 10 | 12 | 14 |
| Daniel | 2* | 13 | 14 |
| Katie | 2* | 2* | 4 |
| Keith | 9 | 9 | 10 |
| Neal | 7 | 10 | 11 |
| Median | 9 | 11 | 11.5 |

*Note:* Phase one and two scores were measured from classroom observations and written summaries using the rubric shown in Table 4.3. The total science practice score was measured based on data collected from classroom observations, Phase 1 summaries, Phase 2 summaries, oral presentations, and journal entries.
The range of scores for each column is 0–14.
There is a statistically significant increase in students' learning of science practices ($z = 2.533$, $r = 0.844$).
* Indicates student did not complete the summary; these students were excluded from the statistical analysis.

alternative energy source (biofuel), algae are good for the environment, and algae can be used for scientific research and development. Incorrect responses, "I don't know," and "No answer," were coded together.

The frequency of these responses on the pre-and posttest is shown in Table 4.5 as a percentage of the total responses. There was a decrease in the last category (I don't know, no answer, or incorrect answer) from 40% to 5%. This decrease can be attributed to an increased understanding of algae as a biofuel, algae as photosynthetic organisms, and that algae benefit the environment. Overall, the data suggest an increased ability to explain the role of algae in our environment, including their use as a biofuel and as a photosynthetic organism. Thus, students not only had a better understanding of scientific concepts such as photosynthesis but also demonstrated an increased recognition of the benefits of algae as a potential biofuel.

Second, we wanted to determine if students recognized what they were doing as important to solving an environmental concern. Several students

**TABLE 4.5 Frequency of Pre- and Postresponses to the Question "What Benefits are Obtained From Growing Algae?**

| Theme | Pre | Post |
|---|---|---|
| Good for the environment | 20% | 35% |
| Alternative energy (biofuel) | 20% | 40% |
| Photosynthesis | 13% | 20% |
| Research & Development | 7% | 0% |
| I don't know/No answer/Incorrect | 40% | 5% |

made specific reference to the idea that what they were doing was important for the environment and the betterment of society. For example, Keith stated, "[This experience] made me feel really good, I felt more important somehow. Like, our data could help change the environment... It's really, like, exciting because what if in the future this is our source of energy. And I can say I was part of it." In this statement he is making a strong connection to his involvement in the research project and the value that he offers to the ABFRG, as well as making a difference with regard to the environment. Another student, Jessie, stated,

> I think that all of our data is useful information because they might want to put algae in wastewater, which has 100% ammonia like Professor Berber said. And it's good to look at different concentrations of ammonia and nitrate to see which concentration the algae best grows in. I think all of our data is probably going to be analyzed by Professor Berber's group... Like, I found it really interesting, to me, the fact that we're using, that we are working with the University helping the environment and find an alternate source of fuel.

Overall, 1 of the 12 students in this class made specific references to the significance of what they were doing with regard to the environment. Collectively, the qualitative data support the finding that students perceived their role in the ABFP as one of service learning as they recognized that what they were doing was of significance to the scientific research community and that they were doing something to better the environment.

Professor Berber saw her involvement as an opportunity for her students to engage in service learning. She has a long history of engaging her students in educational outreach projects, including developing curriculum materials and running professional development workshops for teachers. More recently, Professor Berber has incorporated service learning as a major part of the capstone course she teaches to undergraduate environmental engineering students. She sees service learning as "not just to do something to help the community but to facilitate learning through the service. So students learn by doing and what they are doing in a real project that serves the community" (Berber, personal communication, August 14, 2014). In this capstone course, her students worked with water and wastewater utilities to help solve actual problems. Her involvement in the ABFP took the concept of service learning and expanded it further. In this case, Katia and Grace not only provided service to the high school students and their teacher while gaining the knowledge and skills to facilitate similar projects, but they also profited from the high school students' research. She told us that Katia and Grace "had done aspects of those things in the lab by pulling this project together for the high school students, they were able to do certain tasks that they hadn't been able to do without assistance" (Berber, personal communication, August 14, 2014). That is because the high school students, similar to the capstone students, were given the opportunity to help solve some of Katia and Grace's research problems to test out what variables are important to consider when identifying the best conditions for algae production.

In summary, the ABFP as a school-based service learning project leads to two important findings. First, different members can contribute different types of service to a service learning project. For example, the university researchers perceived their role as contributing to high-quality science for marginalized students at an urban high school while the students perceived what they were doing as a way to help the environment. A second consideration is that service learning can take place within a formal education setting. We believe these two findings came about in part by conceptualizing the ABFP as a contextually based authentic science experience as a collaborative endeavor among university researchers, high school students, and secondary science educators. We discuss the implications of these findings in the next section.

Students believed their experience to be one of authentic science. This is significant because it is not enough to assume that students' perception of the experience will mirror the intention of the education and science researchers designing the intervention. In addition, students demonstrated the learning of scientific content and practices and strengthened their perceptions regarding their research skills as it relates to the ABFP. Thus, a school-based service learning experience is one way to provide high-quality science education for students. Because all students in this study possessed at least one marginalizing factor (ethnicity, SES, gender, ELL, SE) with regard to their science education, we posit that marginalized students can become more engaged in science, improve their learning of science, and develop their sense of civic responsibility when they participate in this type of research and service learning experience.

## DISCUSSION

Four significant findings emerged from this study. First, high school students learned science content and practices by participating in a contextually based authentic science experience. Second, science service learning does not have to be restricted to out-of-school settings as we offer evidence of a successful intervention that takes place in a science classroom. Third, service to the community can have multiple meanings based on the involvement of different stakeholders. For example, Professor Berber saw her involvement as service to the Kindergarten–12 community while students perceived their involvement as service to the scientific community and the community at-large. Fourth, contextually based authentic science experiences that incorporate service learning can help mitigate the effects of marginalization. We address each of these findings in more detail in the following paragraphs.

We demonstrate that students not only learned science content and practices, but they also perceived their experience to be one of authentic science. Students connected their experience to principles of photosynthesis and were able to articulate understandings of these scientific concepts. Students were assessed based on their ability to carry out different science practices, including creating a research hypothesis, designing an experiment, making observations and collecting data, analyzing data, and explaining the results of the research. While the strongest scientific practice was the ability to make observations and collect data, students learned different aspects of research practices laid out by NGSS. An important outcome of the intervention was that students saw their role as working side-by-side with the ABFRG, thus perceiving the ABFRG role as collaborative rather than authoritarian. As a result, students felt empowered in that they were making a significant contribution to the scientific community with regard to the research and globally by attempting to solve an important environmental issue. This is an important feature when designing classroom instruction for students, especially those who have been marginalized. If we want all students to learn science, then they must have the opportunity to do high-quality science that is authentic to themselves and to the science community.

Most service learning, including science, takes place as an out-of-school activity or as a hybrid of in- and out-of-school activities. For example, others have reported service projects in which college students visited local elementary schools to engage students in hands-on science activities (Kamela & Gammon, 2011), in which college students performed well-water testing for local residents (Simmons, 2000), and in which preservice teachers engaged children at a homeless shelter in science (Barton, 2000). An advantage to service learning in an out-of-school setting is that it removes the curricular constraints that teachers must deal with in their classrooms. However, a disadvantage is that it requires teachers, staff, and/or administrators to dedicate additional out-of-school time to the project. The type of school-based service learning project that we reported on here eliminates many of those disadvantages and has the advantage of establishing a tighter connection with the school curriculum.

While different definitions of service learning exist, (i.e., Barton, 2000; Billig, 2000; Simmons, 2000), it is often considered an endeavor blending academic learning with a community need (Anderson, Swick, & Yff, 2001). We offer one way in which service learning can take place within a regular classroom setting so it aligns to the curriculum, allows students to connect their learning to a need in the community, and provides them with a voice. Thus, service learning can be a part of regular classroom instruction if these elements are carefully planned and implemented. Given the demands on teachers to follow state standards, collaboration among all stakeholders is vital to the long-term success of this type of intervention.

Students connected what they were learning to civic responsibility. School-based service learning can motivate students to learn the content by helping them discover how they can make a difference in the community as it can create ownership of and increased engagement in learning. An important aspect of this intervention was incorporating a student reflection component to help them develop heightened awareness of the connection between what they were learning and how they were making a difference.

Service to the community does not have to be limited to the students. By creating an environment in which students collaborate with scientists, the scientists are able to offer service to marginalized K–12 students. When scientists and other professionals are willing to work with marginalized students, they are helping to address an important equity issue by providing students with access to meaningful, high-quality science education. Bencze and Bowen (2009) argue this as a way of providing students with access to necessary cultural and social capital for success in science education.

We offer this intervention for school-based service learning projects in K–12 science classrooms. Combining in-class service learning and authentic science aligned with science standards promotes learning and involvement from all stakeholders, including the students. This model has been shown to work in both formal and informal settings as a way to bring various members of a community together to perform scientific research. In this case, the members included university faculty, teachers, and students to address a prevalent environmental issue.

## NOTE

1. Although the ABFRG was located in a department of environmental engineering, their research resembles what *The Framework for K–12 Science Education* (NRC, 2012) refer to as science rather than engineering practices. Therefore we refer to their research as science, not engineering.

## REFERENCES

Anderson, J., Swick, K., & Yff, J. (2001). *Service learning in teacher education: Enhancing the growth of teachers, their students, and their communities.* Washington, DC: AACTE.

Barton, A. C. (1998). Teaching science with homeless children: Pedagogy, representation, and identity. *Journal of Research in Science Teaching, 35,* 379–394.

Barton, A. C. (2000). Crafting multicultural science education with preservice teachers through service learning. *Journal of Curriculum Studies, 32*(6), 797–820.

Barton, A. C. (2001). Science education in urban settings: Seeking new ways of praxis through critical ethnography. *Journal of Research in Science Teaching, 38*(8), 899–917.

Barton, A. C., & Yang, K. (2000). The culture of power and science education: Learning from Miguel. *Journal of Research in Science Teaching, 37*(8), 871–889.

Basu, S. J., & Barton, A. C. (2007). Developing a sustained interest in science among urban minority youth. *Journal of Research in Science Teaching, 44*(3), 466–489.

Bencze, L., & Bowen, G. M. (2009). A National Science fair: Exhibiting support for the knowledge economy. *International Journal of Science Education, 31*(18), 2459–2483.

Billig, S. (2000). Research on K–12 school-based service learning. *Phi Delta Kappan, 81*(9), 658–664.

Buxton, C. (2006). Creating contextually authentic science in a "low-performing" urban elementary school. *Journal of Research in Science Teaching, 43*(7), 695–721.

Buxton, C., & Lee, O. (2010). *Diversity and equity in science education.* New York, NY: Teachers College Press.

Chapman, A. (2013). *An investigation of the effects of an authentic science experience among urban high school students* (Doctoral dissertation). University of South Florida, Tampa, FL.

Chinn, C., & Malhotra, B. (2002). Epistemologically authentic inquiry for schools: A theoretical framework for evaluating inquiry tasks. *Science Education, 86,* 175–218.

Cohen, J. W. (1988). *Statistical power analysis for the behavioral sciences* (2nd ed.). Hillsdale, NJ: Erlbaum.

Darling-Hammond, L. (1999). America's future: Educating teachers. *Education Digest, 64*(9), 18–35.

Eisenhart, M. (2001). Educational ethnography past, present, and future: Ideas to think with. *Educational Researcher, 30,* 16–27.

Feldman, A., Divoll, K., & Rogan-Klyve, A. (2009). Research education of new scientists: Implications for science teacher education. *Journal of Research in Science Teaching, 46*(4), 442–459.

Feldman, A., Chapman, A., Vernaza-Hernández, V., Ozalp, D., & Alshehri, F. (2012). Inquiry-based science education as multiple outcome interdisciplinary research and learning (MOIRL). *Science Education International, 23*(4), 328–337.

Feldman, A., Divoll, K. A., & Rogan-Klyve, A. (2013). Becoming researchers: The participation of undergraduate and graduate students in scientific research groups. *Science Education, 97*(2), 218–243.

Hsu, P.-L., Roth, W.-M., & Mazumder, A. (2009). Natural pedagogical conversations in high school students' internship. *Journal of Research in Science Teaching, 46*(5), 481–505.

Johnson, R. R. B., & Christensen, L. B. (2013). *Educational research: Quantitative, qualitative, and mixed approaches* (5th ed.). Thousand Oaks, CA: Sage.

Kamela, M., & Gammon, D. (2011). The mobile science center: Promoting science engagement through service learning. *The International Journal of Science in Society, 2*(4), 1–11.

Kardash, C. M. (2000). Evaluation of undergraduate research experience: Perceptions of undergraduate interns and their faculty mentors. *Journal of Educational Psychology, 92*(1), 191–201.

Lave, J., & Wenger, E. (1991). *Situated learning: Legitimate peripheral participation.* New York, NY: Cambridge University Press.

Leech, N., & Onwuegbuzie, A. (2009). A typology of mixed methods research designs. *Qualitative Quantitative, 43,* 265–275.

Miles, M., & Huberman, M. (1994). *Qualitative data analysis* (2nd ed.). Thousand Oaks, CA: Sage.

Morrell, P. D., & Carroll, J. B. (2003). An extended examination of preservice elementary teachers' science teaching self-efficacy. *School and Science Mathematics, 103*(5), 246–251.

National Center for Education Statistics. (2007). *Trends in international mathematics and science studies. 2007 results.* Retrieved from https://nces.ed.gov/nationsreportcard/naepdata/dataset.aspx

National Center for Education Statistics. (2009). *National assessment of educational progress.* Retrieved from http://nces.ed.gov/timss/idetimss/

National Research Council. (2012). *A framework for K–12 science education: Practices, crosscutting concepts, and core ideas.* Washington, DC: The National Academies Press.

NGSS Lead States. (2013). *Next generation science standards: For states, by states.* Washington, DC: National Academies Press.

Oakes, J. (1990). *Multiplying inequalities: The effects of race, social class, and tracking on opportunities to learn mathematics and science.* Santa Monica, CA: Rand Corp.

Oakes, J. (2000). *Course-taking and achievement: Inequalities that endure and change.* Paper presented at the National Institute for Science Education Forum, Detroit, MI.

Patton, M. Q. (2002). *Qualitative research and evaluation methods* (3rd ed.). Thousand Oaks, CA: Sage.

Roberts, J. K., Henson, R. K., Tharp, B. Z., & Moreno, N. (2001). An examination of change in teaching self-efficacy beliefs in science education based on duration of inservice activities. *Journal of Science Teacher Education, 12*(3), 199–213.

Seiler, G. (2001). Reversing the "standard" direction: Science emerging from the lives of African American students. *Journal of Research in Science Teaching, 9*(38), 1000–1014.

Simmons, J. A. (2000). An environmental science approach to service learning in biology. In D. D. Brubaker & J. H. Ostroff (Eds.), *Life, learning, and community: Concepts and models for service learning in biology* (pp. 25–30). Washington, DC: American Association of Higher Education.

Wenger, E. (1998). *Communities of practice: Learning, meaning and identity.* Cambridge, England: Cambridge University Press.

Zacharia, Z., & Barton, A. C. (2004). Urban middle-school students' attitudes toward a defined science. *Science Education, 88*(2), 197–222.

CHAPTER 5

# NURTURING MULTIPLE CURIOSITIES

## Priorities From a National Survey

**Jennifer Foreman**
***Allegheny College***

**E. Jean Gubbins**
***University of Connecticut***

**Cindy Gilson**
***University of North Carolina at Charlotte***

**Micah Bruce-Davis**
***University of Louisiana at Lafayette***

### ABSTRACT

Students from underrepresented groups may avoid science coursework and career aspirations in part due to a perceived disconnect between the intellectual curiosity inherent in scientific disciplines and the empathic need to connect with other humans and enrich their lives. With the creation of many more science, technology, engineering, and mathematics (STEM) focused elementary and high schools, addressing this problematic disconnect seems

*Science and Service Learning*, pages 99–123

vital to recruiting a larger and more diverse cadre of STEM professionals. We do not know whether administrators at these newly formed schools report such practices as priorities in their curricular and instructional decisions. This investigation reports findings of a nationwide survey aimed at determining what priorities exist among administrators at STEM-focused high schools and the implications these priorities may have for integrating service learning and community engagement into science education.

When plague raged through Great Britain in the mid-17th century, Isaac Newton left Cambridge and retreated to his family's country estate where he invented portions of the branch of mathematics we now know as calculus and transformed our scientific understanding of light, color, and planetary motion. What would have happened if Newton had stayed in Cambridge to assist the dying or had turned his brilliant and curious mind toward understanding how plague is spread and how it might be cured? Would the world be better off if he had pursued more practically useful, socially oriented goals? Or was the pursuit of his personal intellectual curiosity more worthwhile in the long run?

Science could perhaps be concisely defined as systematic curiosity. In his book *Curious: The Desire to Know and Why Your Future Depends on It* (2014), Ian Leslie distinguished among three types of curiosity: diversive, epistemic, and empathic. Diversive curiosity is a superficial quest for novelty and excitement, such as what we might engage in when flipping channels or watching Twitter feeds. Epistemic curiosity runs deeper; it is when someone systematically engages in a "wide-ranging desire for intellectual and cultural exploration" (p. xx). Finally, empathic curiosity involves the desire to know what is in the minds and experiences of others and, presumably, to use that knowledge to make others' lives less painful and more beautiful. Epistemic curiosity (such as Newton's) and empathic curiosity (such as Mother Teresa's) can both make lives better for others. Students, however, often perceive these types of curiosity as being separate and disconnected rather than intertwined in a holistic quest to use the complex human mind to solve problems that effect humans and the world they inhabit.

A search of Google Images for "scientist" will provide a perspective consistent with the way most K–12 students think about careers in science—people in white lab coats holding beakers of colorful liquids and peering through microscopes. Although occasionally pictured with a colleague or two, scientists are mainly presented as working alone with physical materials in a sterile laboratory. They are not helping others. They are not tackling community justice issues. They are taking on head work rather than heart work. This prototype of scientific practice as cool, detached rationality may be attractive to certain students, but for many other young people who aspire to help others in a more tangible way, this conception may prevent the development of interest in pursuing a career in science. Efforts to involve K–12 students in experiences that combine scientific understanding and service to others may help many students to integrate careers that use science into their own sense of identity.

Young students often express views of science focusing on specific content knowledge rather than scientific processes and big ideas. The Next Generation Science Standards (NGSS) include a series of four cross-cutting ideas about the nature of science that students should come to understand across their K–12 education. These include:

- Science is a Way of Knowing
- Scientific Knowledge Assumes an Order and Consistency in Natural Systems
- Science is a Human Endeavor
- Science Addresses Questions About the Natural and Material World (NGSS, 2013)

To help students learn these big ideas, teachers and administrators should consider providing more learning opportunities in which they can contrast scientific forms of knowing with other possible epistemologies, enmesh themselves in the fully human and messy process of scientific inquiry, and connect how scientific theories can be used to improve the lives of people and communities as well as the natural world.

Recent federal education policy has favored increasing STEM education in K–12 schools to create "a greater portion of populace that is better prepared in STEM, and generally more STEM literate" (Lynch, Behrend, Burton, & Means, 2013, p. 2). A number of prestigious and selective STEM academies have existed in the United States, primarily designed to serve high-ability students preparing to become future intellectual leaders in STEM fields. Several estimates suggested that between 100 (Subotnik, Tai, Rickoff, & Almarode, 2010) to 315 (Means, Confrey, House, & Bhanot, 2008) STEM-focused K–12 schools are already established. However, even students who do not go on to pursue advanced degrees or STEM careers need a better understanding of these disciplines to become informed, productive citizens (NRC, 2011). To support this goal, the President's Council of Advisors on Science and Technology (PCAST, 2010) recommended creating 1,000 new STEM schools (800 elementary and 200 high schools) by the year 2020. Many of these new schools are designed to be inclusive schools with no achievement or ability-related selection criteria.

In inclusive STEM schools, students' latent talents and interests may emerge when they have opportunities to interact with engaging and enjoyable STEM curricula; a process that has been described as emerging "opportunity structures" (Lynch et al., 2013, p. 4). Opportunity structures in this context refer to pathways for "students underrepresented in STEM to move into rewarding STEM fields" (Peters-Burton, Lynch, Behrend, & Means, 2014, p. 66). Prior research on inclusive STEM schools points to the need for building multiple forms of social capital to provide the opportunity structures for increased participation of females students and students of color to pursue further STEM learning and careers. Peters-Burton and colleagues (2014, p. 66) noted, "structural social capital is built through the collective action of the community." Although these researchers referred mainly to the community within a school, the concepts can be applied more broadly across larger communities in which students and their teachers and families participate. Service learning and community-engaged activities can support broad-based structural social capital, providing opportunity structures by building students' empathic curiosity and desire to participate in science-related community activities.

Service learning and community-engaged learning experiences are distinguished from community service in that the former are intended to reinforce and explicitly link to academic objectives. Most frequently studied in higher education, service learning pedagogies are intended to serve at least four educational objectives: "enhancement of disciplinary-based competencies, development of social and personal responsibility, fostering intercultural competencies, and civic engagement" (Brabant & Braid, 2009, p. 64). Service learning and community-engaged activities are among a small set of "high impact" educational practices found to increase undergraduate students' academic achievement and long-term success (Kuh, 2008), particularly for students most at risk of low achievement or noncompletion.

One model that has been suggested for providing escalating levels of community engaged learning in K–12 education is the Operation Houndstooth Intervention Theory (Renzulli, Koehler, & Fogarty, 2006). This model seeks to develop social capital by promoting student character traits, including the concepts of epistemic curiosity (romance with a discipline) and empathic curiosity (sensitivity to the concerns of others). The interventions range from a "Rally Round the Flag" approach (such as promoting character virtues through visual displays in the school) to "Direct Involvement" activities. "Direct Involvement I" activities give students opportunities to participate in service learning or community engagement. "Direct Involvement II" is a form of Type III enrichment triad model activities that require students to create a new socially beneficial product or service rather than simply participating in an existing one. Some empirical studies have recently demonstrated positive outcomes of these interventions (Reilly, 2009; Sands, 2012).

## INVESTIGATION USING A NATIONAL SURVEY

Given the promise of service learning and community engaged opportunities in science education to spark the interest of underrepresented students, it remains unclear whether teachers and administrators favor these practices. With the recent creation of many STEM-focused K–12 schools following the PCAST (2010) recommendation, we know little about which curricular and instructional practices are considered priorities for administrators at STEM-focused schools. To better understand these issues, we addressed the following questions across a national sample of STEM-focused high schools:

1. What are teachers and administrators' perceptions of the importance and frequency of various curricular and instructional practices in STEM-focused high schools?
2. How do teachers and administrators view curricular and instructional practices consistent with building epistemic and empathic curiosity?

We conducted an extensive search to identify the sampling frame of STEM high schools throughout the United States. We identified a total of 949 unique STEM schools by searching websites, reviewing articles identified through electronic searches using key search terms, contacting state departments of education, and soliciting names of schools from the National Consortium of Specialized Secondary Schools of Mathematics, Science, and Technology. These schools served as the sample for the study.

To develop the STEM administrator and STEM teacher survey for the study, we first observed and interviewed key stakeholders (teachers, students, and administrators) in 12 STEM high schools. Using the qualitative data from these visits, we inductively coded a comprehensive set of curricular and instructional strategies and practices, school policies, and school culture factors that were present at the STEM schools (Bruce-Davis et al., 2014). Common features across schools, especially those that seemed to represent best practices in STEM education based on the literature, were developed into item stems for the surveys.

The 48-item administrators' survey was divided into five sections: Professional Culture, Curricular and Instructional Practices, Policies and Procedures, Description of Practices, and Demographics. In the Professional Culture and Curricular as well as the Instructional Practices sections, one response scale focused on administrators' perceived importance of each item; the other scale assessed the frequency with which administrators perceived that the practice occurred. Only the importance of each item was measured in Policies and Procedures. The Description of Practices section contained four open-ended questions regarding faculty policies and program evaluation. Finally, the Demographics section included questions about structural features of the school as well as admissions criteria, course offerings, and students' eligibility for free and reduced lunch.

Importance was measured on a 6-point Likert-type scale with anchors ranging from "unimportant" (1) to "essential" (6). The Frequency scale response options were (1) "never," (2) "once or twice a year," (3) "once or twice a grading period," (4) "once or twice a month," (5) "at least once a week," (6), and "every day" (6). Both scales included a not applicable option.

The 41-item STEM teacher survey was composed of four sections: School Climate, Curricular Approaches, Instructional Strategies, and Learning Environment. The first three sections of the teacher survey used both the Importance and Frequency scales, but only the Importance scale was used in the final section.

## SAMPLE

We sampled all 949 identified schools by sending print and electronic versions (via email links) of the survey to each site. Administrators were asked to forward copies of the teachers' surveys to five of their faculty members. Of those identified, 205 administrators and 777 teachers completed the surveys. Teachers from 291 unique schools completed the surveys, indicating teachers at some schools chose to respond even though their administrators did not. Respondents were not asked for personally identifying information, but were asked to supply the state in which their school operates. The geographic distributions for the administrators and teachers who responded to the survey are shown in Figures 5.1 and 5.2.

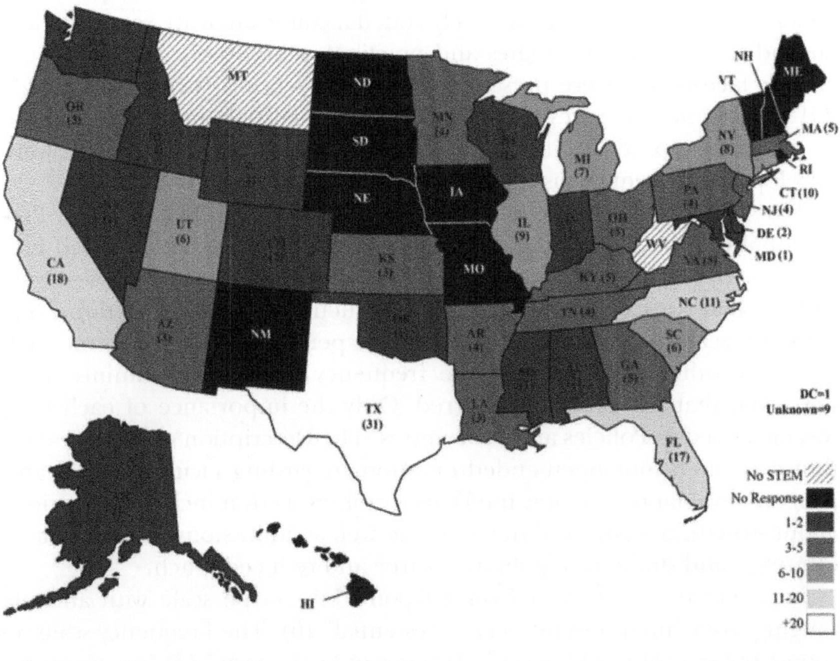


**Figure 5.1** Geographic distribution of STEM-focused high school administrator survey respondents.

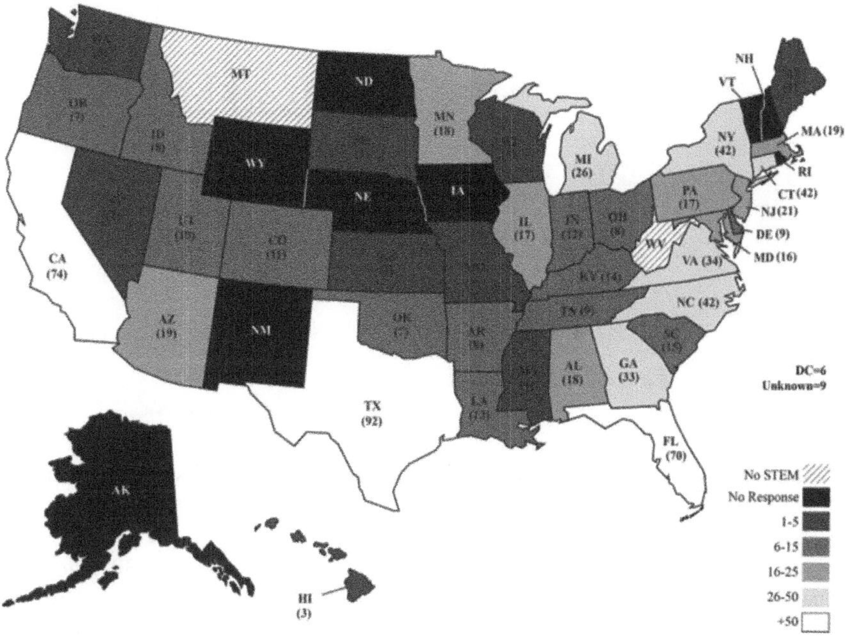


**Figure 5.2** Figure 2. Geographic distribution of STEM-focused high school teacher survey respondents.

## School and Student Characteristics

### *School Structure and Demographics*

Of the 205 administrator respondents, the largest number reported that they served a student population from urban areas (38.0%), followed by suburban (31.2%), and then rural areas (19.0%). Some administrators (7.8%) reported serving students from multiple urbanicities. The administrator respondents identified a wide variety of school types. As indicated in Figure 5.3, the greatest percentage (25.9%) classified their school as a comprehensive high school; sizeable numbers categorized their school as a magnet school (19.5%) or as a charter school (16.6%). Nearly 10% of the administrators selected multiple school types to describe their schools. The vast majority (86.8%) of respondents described their school program's organization as full day.

### *School Size*

Approximately half of the administrators (50.2%) reported schools serving more than 400 students; approximately one third (34.1%) reported between 200 and 400 students, and 12.7% reported serving fewer than 200 students. Nearly half of the schools in the sample (48.2%) employed fewer

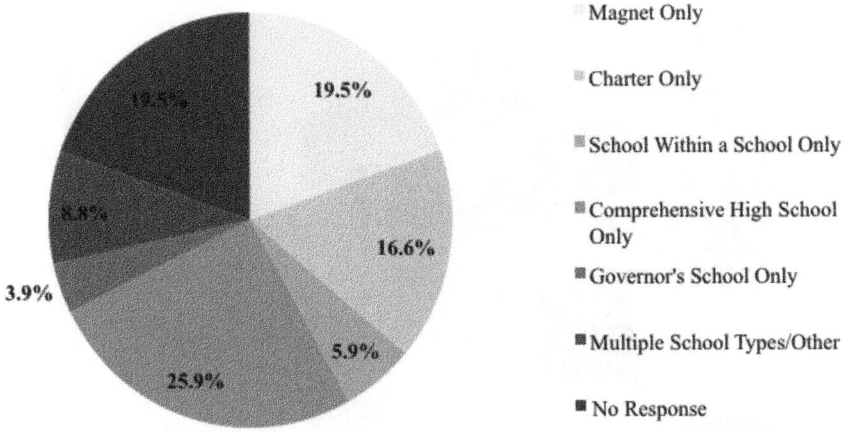


**Figure 5.3** Percent of STEM-focused high school administrator survey respondents who selected each school type.

than 30 teachers and the other half (51.2%) had more than 30 teachers. Slightly more than half (60.3%) had three or fewer administrators, while in a sizeable minority (39.7%) more than three administrators were on staff.

***Admission Criteria***

Administrators were allowed to select more than one response when indicating criteria used for admission to their schools. The most frequently reported criterion used to select students (45.4%) was students' grades, report cards, or transcripts. More than a third of respondents (40.0%) reported using no admission criteria for student enrollment, indicating that these schools were inclusive STEM-focused high schools. Figure 5.4 summarizes the responses to this item. Other admission criteria added by respondents included lottery systems, auditions, and residency requirements.

***Course Offerings***

Nearly half of the respondents' schools offered more than 10 science and mathematics courses per year, while only a few offered more than 10 technology or engineering courses (see Table 5.1).

***Student Characteristics***

About 28% of administrators in the sample reported serving a student population with less than 25% eligibility for free and reduced lunch; approximately the same proportion (27%) of respondents were from schools with 26% to 50% student eligibility. Slightly fewer administrators reported serving populations with 51% to 75% eligibility, and even fewer served populations among the highest category of eligibility (76% to 100%). Despite

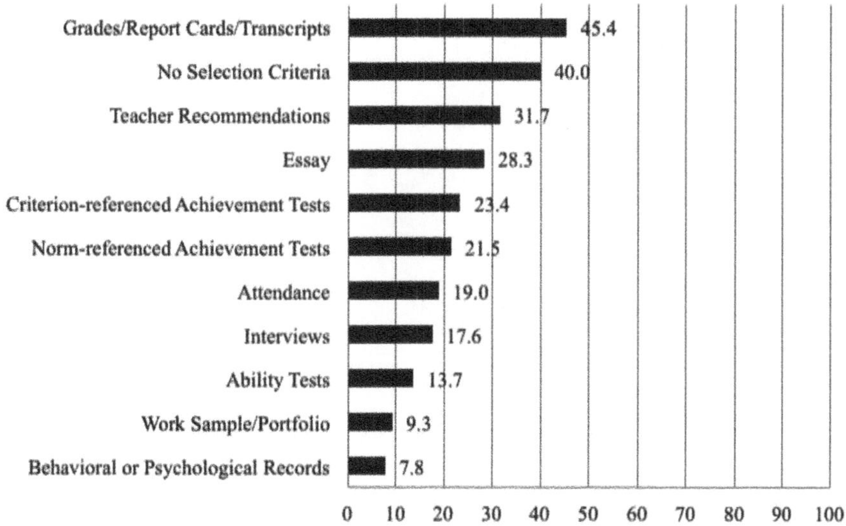


**Figure 5.4** Percent of STEM-focused high schools using different selection criteria for student admittance.

**TABLE 5.1 Percentage of Administrator Responses to Different Ranges of Courses Offered Each Year in the Four STEM Disciplines**

| Range of Courses | Science | Technology | Engineering | Mathematics |
|---|---|---|---|---|
| Fewer than 10 courses | 49.8 | 74.1 | 82.0 | 50.7 |
| Between 10 and 20 courses | 32.7 | 14.6 | 4.8 | 32.2 |
| More than 20 courses | 10.7 | 2.4 | 0.0 | 9.2 |
| No response | 6.8 | 8.8 | 12.7 | 7.8 |

these differences, a full range of school-level SES levels was represented in the schools responding to the survey (see Figure 5.5).

Student gender ratios were well balanced at nearly all of the schools surveyed. A small percentage (4.5%) of schools served fewer than 25% or greater than 75% female students. The survey did not ask administrators if their school was a single-sex school, which may have been the case for these few schools.

## ANALYSES OF ADMINISTRATORS AND TEACHERS' SURVEY RESULTS

We used descriptive statistics to summarize administrators and teachers' responses to the survey items. Medians were selected because of the ordinal

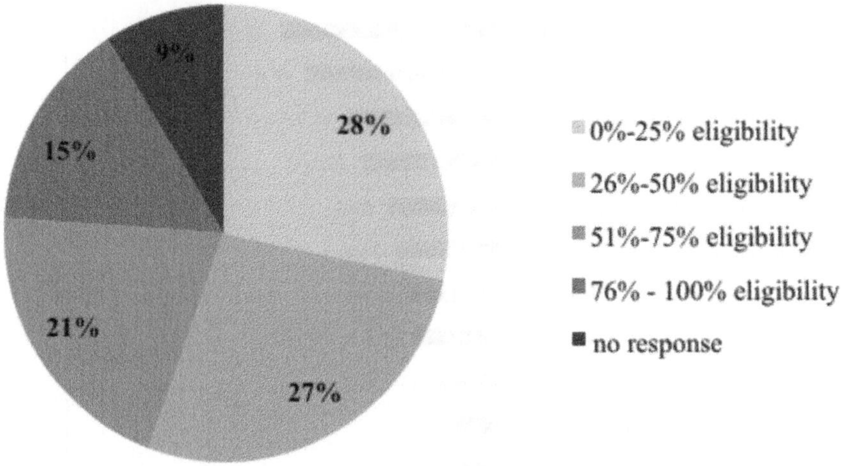


**Figure 5.5** Percent of students eligible for free and reduced priced meals in responding STEM-focused high schools.

nature of the data and the highly skewed response patterns for many items on the Importance scale (Hinkle, Wiersma, & Jurs, 2003). We reported these statistics for respondents' choices on both scales of the survey items. We also examined items for which there might be discrepancies between respondents' perceived importance and reported frequency of use by calculating the difference in medians. We considered more than one point of difference between the medians on the two scales to be indicative of a potential mismatch between perceived and enacted priorities. For some items, this discrepancy was not problematic due to high importance practices that would not be expected to occur frequently. But for many other items, we believed it was important to compare the differences between responses on the two scales. As another way to describe the relative importance of the items on the survey, we reported the percentage of respondents who selected one of the two highest possible importance scale rating categories ("very important" or "essential") for each item.

## ADMINISTRATORS' SURVEY RESULTS

### Professional Culture

Two items, "Encourage teachers to ask open-ended questions with no solution path" and "Provide students with opportunities to participate in extracurricular activities," received a median importance rating of "essential" by the administrators who responded (see Table 5.2). No Professional Culture items received a median frequency rating of "every day"; however,

**TABLE 5.2 Administrators' Median Responses to the Professional Culture Items (*n* = 205)**

| Item | Median Importance | Median Frequency | Median Difference |
|---|---|---|---|
| 1. Communicate a STEM-specific vision of the school. | 5 | 4 | 1 |
| 2. Provide scheduled time for teacher collaboration within each STEM discipline. | 5 | 4 | 1 |
| 3. Provide scheduled times for teacher collaboration across STEM disciplines. | 5 | 4 | 1 |
| 4. Provide scheduled times for teacher collaboration between STEM and non-STEM disciplines. | 4 | 3 | 1 |
| 5. Promote change through faculty involvement in decision-making. | 5 | 4 | 1 |
| 6. Provide sustained opportunities for teacher learning within the school. | 5 | 4 | 1 |
| 7. Allow teacher flexibility in modifying curriculum. | 5 | 5 | 0 |
| 8. Conduct observations of teachers focused on their use of inquiry-based pedagogy. | 5 | 4 | 1 |
| 9. Encourage teachers to ask open-ended questions with no single answer or solution path. | 6 | 5 | 1 |
| 10. Recruit students from culturally diverse or underrepresented groups. | 5 | 3 | 2 |
| 11. Provide students with opportunities to participate in extracurricular activities. | 6 | 5 | 1 |
| 12. Require students to complete community service. | 5 | 4 | 1 |
| 13. Provide counseling services for students' social-emotional needs. | 5 | 5 | 0 |
| 14. Provide specialized counseling for students' long-term career plans. | 5 | 4 | 1 |

the following items had a median rating of occurring "at least once a week": "Allow teachers flexibility in modifying curriculum"; "Encourage teachers to ask open-ended questions with no single answer of solution path"; Provide students with opportunities to participate in extracurricular activities"; and "Provide specialized counseling services for students' social-emotional needs" The only item on the Professional Culture section with a greater than one point difference in medians between the two response scales was "Recruit students from culturally diverse or underrepresented groups (e.g., females, African Americans, Native Americans, low SES)." Although administrators viewed this practice as very important, it only occurred on average once or twice a grading period, which may reflect their recruitment and admissions cycles.

Although STEM administrators generally rated most items in the Professional Culture section highly, differences existed among items in the percentage of administrators who selected either very important or essential. Over 80% of administrators rated within-field teacher collaboration and sustained professional development as "very important" or "essential." In contrast, less than 50% of administrators rated "Collaboration between STEM and non-STEM teachers" as very important or essential. Figure 5.6 shows the percentage of administrators rating each Professional Culture item as "very important" or "essential."

## Curricular and Instructional Practices

Providing students with access to basic STEM resources, such as computers, Internet, graphing calculators, and laboratory equipment received a median importance rating of essential. All other items in this section had a median rating of either "important" or "very important." On the Frequency scale, administrators reported providing access to basic STEM resources

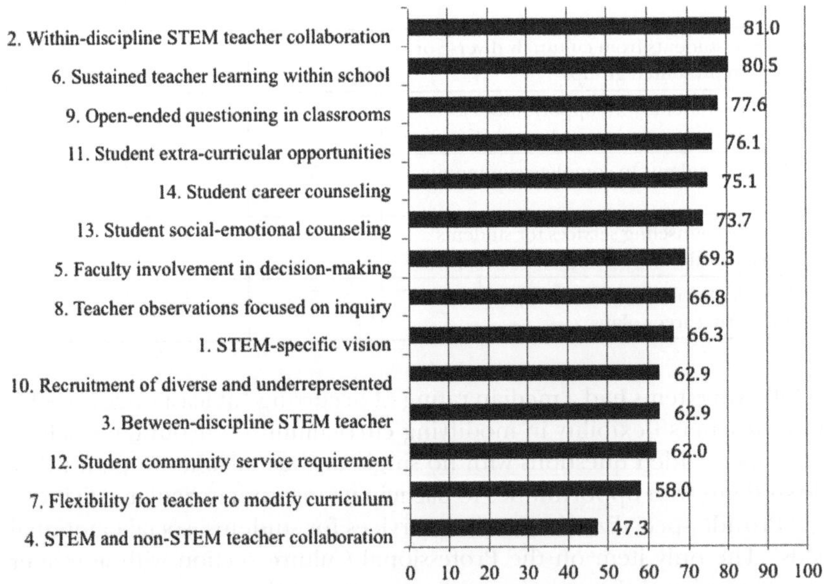


**Figure 5.6** Percent of STEM-focused high school administrator respondents who rated each survey item as "very important" or "essential."

and providing access to visual and performing arts classes with a median frequency of "every day." Items associated with program evaluation, faculty research, student participation in STEM competitions, and student shadowing of STEM professional had a median frequency rating of "once or twice a year." Approximately one third of the administrators reported "never" in response to requiring external program evaluation and promoting student participation in international STEM competitions. A number of the items in this section had a difference in median importance and frequency ratings, including internal program evaluation, student participation in national STEM competitions, and student shadowing of STEM professionals. Some discrepancies between importance and frequency ratings, such as participation in competitions, may be due to factors outside of the administrators' control. Results of the curricular and instructional practices section of the survey are summarized in Table 5.3. Figure 5.7 illustrates the percentage of administrators who responded to each of the curricular and instructional practices items with a rating of either "very important" or "essential." Providing basic STEM resources and requiring summative and formative assessment of students received the highest percentage of these ratings, while only about 25% of respondents rated promoting faculty research and requiring external program evaluation as very important or essential.

***Policies and Procedures***

Only the Importance scale was rated in the Policies and Procedures section. Administrators gave a median importance rating of "very important" to all of the items in this section except for the item referring to providing students opportunities to obtain industry certifications, which had a median rating of "important" (Figure 5.8).

Analysis of the open-ended questions in this section of the survey revealed that while there was general consensus among administrators about teachers' involvement in the processes surrounding developing curriculum, the degree of reported involvement varied. Though a small group of administrators stated that teachers had little autonomy over the curricula and courses offered, most perceived them as essential to the development of courses if relevant and challenging work was offered for their students. For example, one administrator stated the following:

> The staff is very involved by meeting on a regular basis to discuss the needs as well as the current trends of STEM. STEM is a constantly evolving curriculum especially in the area of technology and it is important to stay current so that we can meet the instructional needs of our students. Technology changes almost daily and it is important that our curriculum moves with it.

The reported amount of time devoted to examining trends in STEM for curricular integration in meetings was highly variable. For some schools,

**TABLE 5.3 Administrator Responses to the Curricular and Instructional Practices Items (*n* = 205)**

| Item1 | Median Importance | Median Frequency | Median Difference |
|---|---|---|---|
| 15. Require preassessment of student knowledge in STEM classes | 4 | 3 | 1 |
| 16. Require formative evaluation of student progress | 5 | 5 | 0 |
| 17. Require summative evaluation of student progress | 5 | 4 | 1 |
| 18. Require that the effectiveness of the specialized STEM program be internally assessed | 5 | 2 | 3 |
| 19. Require that the effectiveness of the specialized STEM program be externally evaluated | 4 | 2 | 2 |
| 20. Provide time for students to meet with research advisors | 5 | 3 | 2 |
| 21. Promote faculty-based research | 4 | 2 | 2 |
| 22. Provide students access to professional STEM journals | 4 | 3 | 1 |
| 23. Provide students access to basic STEM resources | 6 | 6 | 0 |
| 24. Provide students access to advanced STEM resources | 5 | 5 | 0 |
| 25. Promote student involvement in national STEM competitions | 5 | 2 | 3 |
| 26. Promote student involvement in international STEM competitions | 4 | 2 | 2 |
| 27. Provide students access to visual and performing arts classes | 5 | 6 | -1 |
| 28. Provide students opportunities to shadow professionals in STEM fields | 5 | 2 | 3 |
| 29. Offer students dual enrollment at local colleges or universities | 5 | 5 | 0 |
| 30. Provide students opportunities to complete internships in STEM fields | 5 | 3 | 2 |

teacher involvement was restricted to in-service sessions or occasional meetings; others scheduled weekly meetings where teachers and peers developed curricula collaboratively. One administrator stated, “We spent a week last summer developing schoolwide learning outcomes.”

When administrators were asked to describe the qualities and credentials they considered when hiring new teachers, they reported a wide range of criteria with little consensus as to what traits were most important. Many

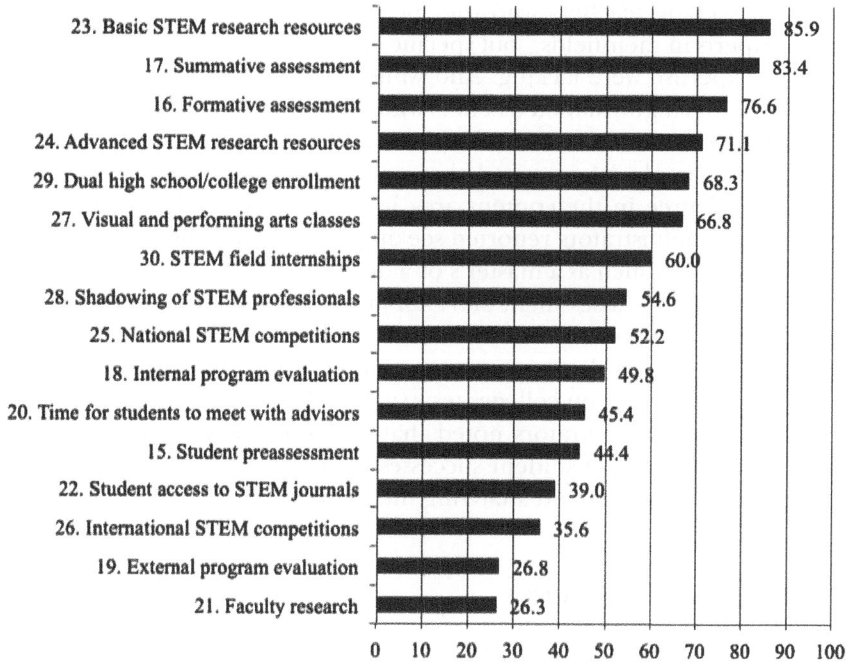


**Figure 5.7** Percent of STEM-focused high school administrator respondents who rated each survey item as "very important" or "essential."

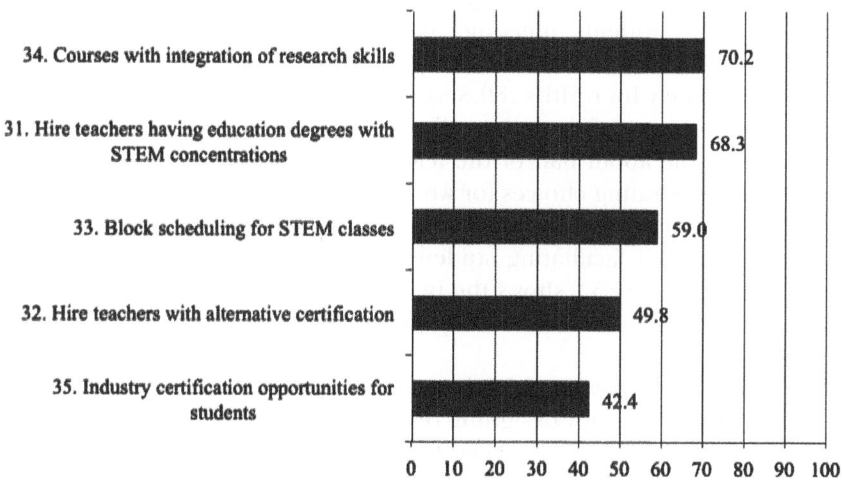


**Figure 5.8** Percent of STEM-focused high school administrator respondents who rated each survey item as "very important" or "essential."

reported seeking teachers who they perceived were passionate and who were "experts in their fields," but specific defining features of what they felt defined expertise were lacking. And while many statements revealed that administrators highly valued teachers who had "knowledge of the content," the most-desired criteria seemed to be credential related. Survey respondents, for example, often stated they expected faculty hires to possess a bachelor's degree in the content area in which they taught, and a small number of administrators reported seeking teachers with higher-level content-area degrees such as a master's or a doctorate.

In discussing the assessment of STEM program curricula, administrators report using standardized test scores as the means for determining the success of their courses. For example, one administrator stated, "We review student grades, scores on colleges tests such as the PSAT, SAT, ACT, and AP exams." Some administrators noted that they integrated data from classroom observations and student successes in research competitions; and a few reported collecting and analyzing their own data.

## Teachers' Survey Results

### *School Climate*

The median teacher importance rating for offering students tutoring or extra help in STEM courses and encouraging students to use their knowledge to better the world was "essential." All other items in this section obtained a median importance rating of "very important." Two of the School Climate items rated as "very important but only enacted once or twice a year related to students' meeting" and "collaborating with STEM professionals." Ratings for the School Climate items are summarized in Table 5.4.

Approximately four fifths (80.8%) of the teacher respondents reported it was "very important" or "essential" to offer students extra academic help if needed, while about half of the teachers responded with these highest two importance rating choices for working to enhance their schools' reputations, providing opportunities for students to meet with diverse STEM professionals, and facilitating student collaboration with working STEM professionals. Figure 5.9 shows the percentage of teachers who rated each of the School Climate items as "very important" or "essential."

### *Curriculum Development and Implementation*

The median rating for designing real-world-oriented curriculum was "essential," with most other items receiving median ratings of "very important" on the Importance scale. Several items, including encouraging students to present products to authentic audiences, were rated as very important but only reported as occurring in practice infrequently (see Table 5.5).

**TABLE 5.4 Teachers' Median Responses to School Climate Items (*n* = 777)**

| Item | Median Importance | Median Frequency | Median Difference |
|---|---|---|---|
| 1. Work to enhance and promote the reputation of excellence at your school | 5 | 3 | 2 |
| 2. Collaborate in STEM curriculum development | 5 | 3 | 2 |
| 3. Arrange collaborative projects for students with working professionals | 5 | 3 | 2 |
| 4. Expect students to maintain a professional lab | 5 | 5 | 0 |
| 5. Offer students tutoring/extra help in STEM classes if needed | 6 | 5 | 1 |
| 6. Offer guidance and counseling for student social/emotional needs | 5 | 5 | 0 |
| 7. Celebrate student accomplishments, achievements, and awards | 5 | 4 | 1 |
| 8. Encourage students to use their knowledge for the betterment of the world | 6 | 5 | 1 |
| 9. Offer students opportunities to meet with STEM professionals of various backgrounds | 5 | 2 | 3 |

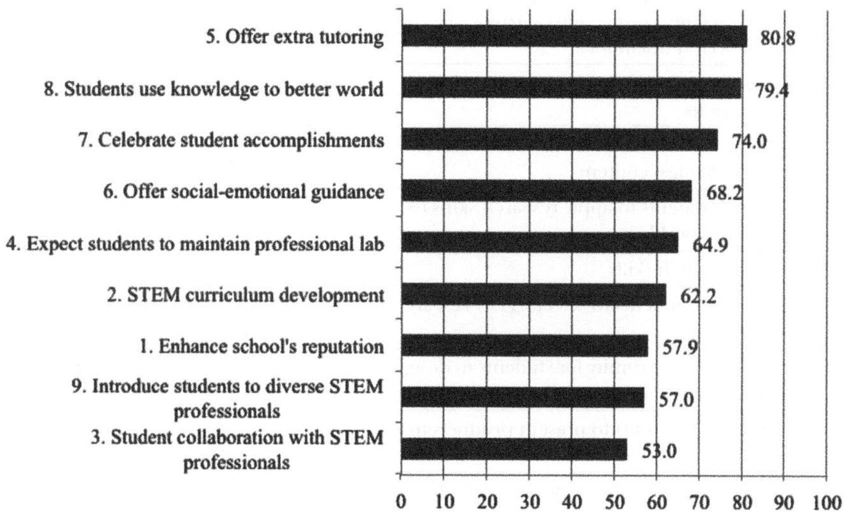


**Figure 5.9** Percent of STEM-focused high school teacher respondents who rated each survey item as "very important" or "essential."

Teachers rated most of the items in the Curricular Approaches section as "very important" or "essential." Few (21.6%) rated adopting curricular units without modification in these highest two categories of importance. For the other items, the percentage ranged from approximately 50% to 80% of teacher respondents (Figure 5.10).

**TABLE 5.5 Teachers' Median Responses to Curricular Approaches Items (*n* = 777)**

| Item | Median Importance | Median Frequency | Median Difference |
|---|---|---|---|
| 10. Teach academic writing skills | 5 | 5 | 0 |
| 11. Adopt preexisting challenging and advanced STEM units of study without making modifications | 4 | 3 | 1 |
| 12. Modify preexisting challenging and advanced STEM units of study | 5 | 4 | 1 |
| 13. Create challenging and advanced STEM units of study | 5 | 4 | 1 |
| 14. Modify units to meet students' readiness levels | 5 | 5 | 0 |
| 15. Emphasize depth of conceptual understanding of STEM topics | 5 | 5 | 0 |
| 16. Design problem-based learning opportunities | 5 | 5 | 0 |
| 17. Model making connections across and within STEM disciplines | 5 | 4 | 1 |
| 18. Design curriculum that promotes real-world applications | 6 | 5 | 1 |
| 19. Integrate controversial and/or timely STEM topics into class content | 5 | 4 | 1 |
| 20. Require students to apply research skills to real-world problems | 5 | 4 | 1 |
| 21. Teach research skills | 5 | 4 | 1 |
| 22. Encourage students to select STEM research topics | 5 | 3 | 2 |
| 23. Provide an opportunity for students to design and complete self-selected research project(s) | 5 | 3 | 2 |
| 24. Encourage students to present products to authentic audiences | 5 | 3 | 2 |
| 25. Provide explicit lessons to teach students to take notes effectively | 5 | 3 | 2 |
| 26. Provide direct instruction to students on time management skills | 5 | 3 | 2 |

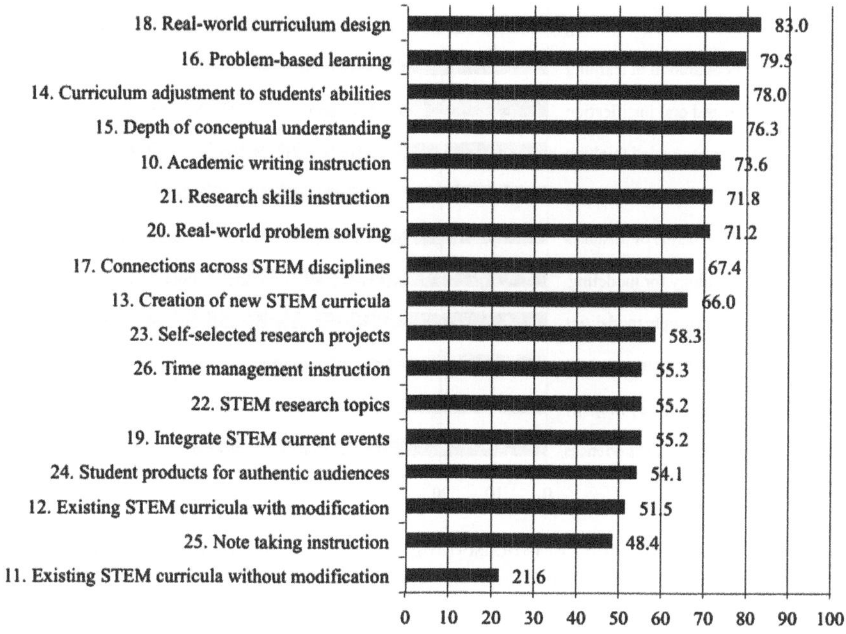


**Figure 5.10** Percent of STEM-focused high school teacher respondents who rated each survey item as "very important" or "essential."

***Instructional Strategies***

Teachers rated all items in the Instructional Strategies section as quite important, with all items receiving a median rating of either "very important" or "essential." In terms of frequency, nearly all of the items in the section had a median frequency of "at least once a week," except for encouraging student questioning, which had a median frequency rating of "every day." No items in this section had a difference between the median importance and the median frequency of greater than one point. The largest percentage of teachers rated student questioning as "very important" or "essential." This information is summarized in Figure 5.11.

***Learning Environment***

In the section on Learning Environment, teachers only selected responses from the Importance scale options. The median rating for promoting a common vision of excellence at the school was "essential," and directly instructing students on help-seeking behaviors was "very important." The vast majority of teachers (83.8%) considered promoting a common vision of excellence for the school to be "very important" or "essential," with a sizeable majority (70.7%) responding with one of the two highest importance

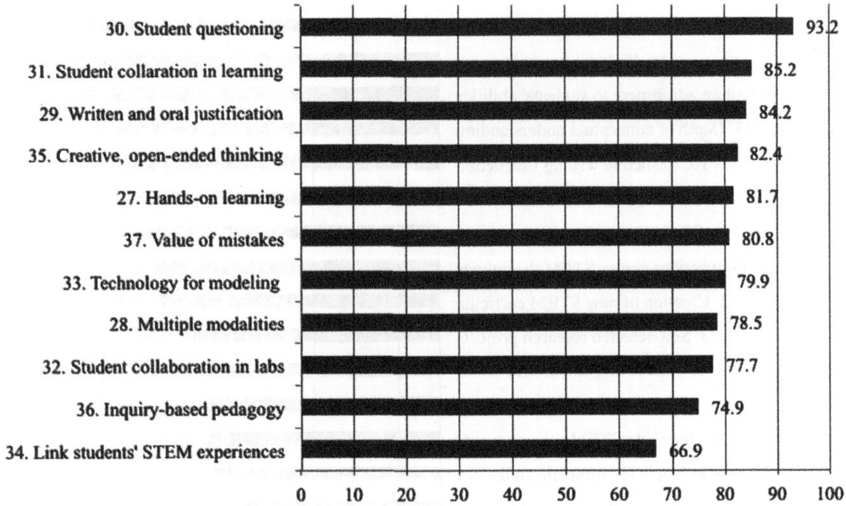


**Figure 5.11** Percent of STEM-focused high school teacher respondents who rated each survey item as "very important" or "essential."

ratings for providing direct instruction in seeking help. In contrast, fewer than half of the teacher respondents perceived promoting STEM careers over other careers (46.2%) or student articulation of long-term career plans (39.4%) to be "very important" or "essential."

## SUMMARY AND DISCUSSION

This study offered an exploratory description of the importance and frequency of various policies and practices in STEM-focused high schools in the United States. Items for the surveys were developed based on inductive analyses of common practices found in different types of STEM-focused high schools (e.g., selective, inclusive, and career and technical schools) with varied types of service delivery models (e.g., magnet, charter, school-within-a-school, comprehensive, and Governor's schools). The purpose of the study was to gather a nationally representative sample of administrators and teachers' perceptions about practices and policies to help identify and assess the critical components (Lynch et al., 2013) of STEM-focused education, and especially those practices related to service learning and community-engaged activities. This evaluative process led to the beginning stages of developing a descriptive framework based on the features unique to these specialized academic settings, which might inform future inductive and deductive studies, the creation of new STEM schools, and improvement of those currently operating (Scott, 2012).

Administrators rated three areas as essential: "Provide students access to basic resources needed to engage in STEM-related research and projects (e.g., lab equipment, computers, Internet, graphing calculators, lab supplies)"; "Provide students with opportunities to participate in extracurricular activities"; and "Encourage teachers to ask open-ended questions with no single answer or solution path." Conversely, administrators perceived the items addressing faculty-based research, external program evaluation, and student participation in international STEM competitions to be relatively unimportant. A number of administrator items rated as very important were reported as occurring with low frequency, such as "Recruit students from culturally diverse or underrepresented groups (e.g., females, African Americans, Native Americans, low SES"; "Provide time for students to meet with research advisors"; and "Provide students opportunities to complete internships in STEM fields."

Administrators in our sample also did not place a relatively high priority on requiring students to complete community service, with only 62% rating the item as "very important" or "essential." Although we have distinguished between community service and more academically supportive forms of service learning and community engagement, our survey did not indicate these forms of learning as one of the higher priorities for STEM-focused high school administrators. Two other items that tapped into community engagement—providing opportunities for students to shadow STEM professionals and complete STEM internships—received significantly lower frequency ratings than importance ratings. Modal frequency for both these items was "once or twice a year." Approximately 55% to 60% of administrators viewed these two opportunities for students to interact with the broader community as very important or essential.

Teachers rated importance and frequency quite high for most of the items on the teacher survey. Some 10 items received median importance ratings of "essential" from teacher respondents, 28 items had a "very important" median rating, and only three items had an "important" median rating. Across the four sections, the item with the highest proportion of very important or essential ratings on the STEM teacher survey was "Encourage student questioning." It may indeed be somewhat alarming that 7% of teachers in the survey did not rate student questioning as "very important" or "essential," given the centrality of curiosity to scientific thinking and inquiry and literature on establishing inquiry-based teaching and learning environments (see Saunders-Stewart, Gyles, & Shore, 2012). The least important items to teachers were "Adopt preexisting challenging and advanced STEM curricula without making modifications"; "Promote STEM careers over other career options"; and "Require written or oral articulation of long-term career plans beyond undergraduate education." It is possible that these career-planning tasks do not reflect the missions of a majority of STEM high schools. Perhaps teachers are uncertain of how to provide such support or perhaps teachers feel that placing these unnecessary academic burdens on adolescents is exclusive of their roles as teachers. In either event, this perception is worth noting because of its disparity with the societal goal of increasing the number of students who pursue STEM-related majors in college and enter STEM-related careers.

Several items that related to service learning or community-engaged learning showed similar patterns across teachers in our sample to what we saw from administrators. Arranging collaborative projects among students and working professionals; offering students opportunities to meet STEM professionals of various backgrounds; requiring students to apply research skills to complex real-world problems; and encouraging students to present products to authentic audiences were all rated highly important by teachers, but not as important as most of the other items. The modal frequency of these items was "once or twice a year." However, over 80% of teachers rated "essential" or "very important" teaching their students to use knowledge for the betterment of the world.

## LIMITATIONS

Several limitations influenced the findings of this study. First, the Importance scale appeared to have considerable ceiling effects for many of the items. Although it is not surprising that administrators and teachers would find many practices to be "very important," research should continue to document which features are the most fundamental to high-quality STEM education. In future studies, researchers may obtain more variable responses by scaling responses without affixing anchor words such as important or essential to the scale points.

From the nearly 1,000 high schools to which we distributed surveys, 205 administrators and 777 teachers responded from 291 different schools. Because the response rate was not higher, it is reasonable to consider that those administrators and teachers who chose to respond may not have been fully representative of the entire population. Perhaps administrators and teachers who were most enthusiastic or had the most available time were able to respond, which could create bias in the results. In short, a higher response rate could have strengthened the validity of our findings and increased the generalizability of our recommendations.

## RECOMMENDATIONS AND CONCLUSIONS

As more STEM-focused schools and programs are developed, it is critical to systematically identify them and determine the impact they have on their students. The National Research Council (2011) established three broad goals for K–12 STEM education in the United States: expand the number of students who ultimately pursue advanced degrees and careers in STEM fields; expand the STEM-capable workforce; and increase science literacy for all students. Effective instruction and conditions and cultures that support learning (NRC, 2013) were identified as the two most proximal influences that should enable students to attain these goals. This study's findings indicated that teachers and administrators rated many related survey items as "very important" or "essential," but the frequency of some practices did not match their perceived importance. These differences may exist because of barriers and limitations to implementing these practices in their respective schools, such as lack of time, materials, or teacher training. Future studies could investigate administrators and teachers' perceptions about why the disparity occurs. In addition, the ways that schools address their effectiveness to fulfill their mission should be examined.

Service learning and community-engaged learning activities have the potential to create opportunity structures that will enable underrepresented students to develop and integrate epistemic and empathic curiosity. When students can picture themselves in STEM career roles, they begin to establish an identity and sense of belonging that increases their likelihood of persisting in further STEM learning and careers. What little evidence exists suggests that, while administrators and teachers may recognize the importance of nurturing these through community engaged pedagogies, infrequent opportunities currently exist for students to make tangible connections between their science education and work that benefits their communities. This is true even in schools that are STEM focused in their mission.

Prior federal policy focused on standardized assessment of academic content learning in science may have discouraged many teachers and administrators from pursuing more community engaged experiences for their students. Possibly, the recent passage of the Every Student Succeeds Act (2015) will promote accountability that also measures less traditional aspects of students' outcomes, such as resilience, curiosity, and engagement. If such measures are considered in determining schools' success, then teachers and administrators may feel more freedom to explore service learning opportunities. The Next Generation Science Standards also hold the potential to make science education more integrated into students' holistic understanding of human knowledge construction, bringing a personal dimension to a set of fields that many students may feel is unapproachably dry and detached from human concerns. The nation needs many more young people of all backgrounds and abilities to become excited, engaged, and proficient in scientific thinking. Service learning and community-engaged activities may hold one important key in accomplishing this long-term goal.

## REFERENCES


Brabant, M., & Braid, D. (2009). The devil is in the details: Defining civic engagement. *Journal of Higher Education Outreach and Engagement, 13*(2), 59–88.

Bruce-Davis, M. N., Gubbins, E. J., Gilson, C. M., Villanueva, M., Foreman, J. L., & Rubenstein, L. D. (2014). STEM high school administrators', teachers', and students' perceptions of curricular and instructional strategies and practices. *Journal of Advanced Academics, 25*(3), 272–306.

Every Student Succeeds Act of 2015, PL 114-95.

Hinkle, D. E., Wiersma, W., & Jurs, S. G. (2003). *Applied statistics for the behavioral sciences* (5th ed.). Boston, MA: Houghton Mifflin.

Kuh, G. D. (2008). Excerpt from high-impact educational practices: What they are, who has access to them, and why they matter. Washington, DC: AACU.

Leslie, I. (2014). *Curious: The desire to know and why your future depends on it.* New York, NY: Basic Books.

Lynch, S. J., Behrend, T., Peters-Burton, E., & Means, B. (2013, April 6–9). *Inclusive STEM-focused high schools: STEM education policy and opportunity structures.* Presentation to the NARST 2013 Annual International Conference, Puerto Rico. Retrieved from https://ospri.research.gwu.edu/sites/ospri.research.gwu.edu/files/downloads/Inclusive%20STEM-focused%20High%20Schools-%20%20STEM%20Education%20Policy%20and%20Opportunity%20Structures.pdf

Means, B., Confrey, J., House, A., & Bhanot, R. (2008). *STEM high schools: Specialized science technology engineering and mathematics secondary schools in the US.* Menlo Park, CA: SRI International.

National Research Council (NRC). (2011). *Successful K–12 STEM education: Identifying effective approaches in science, technology, engineering, and mathematics.* Retrieved from http://www.stemreports.com/wp-content/uploads/2011/06/NRC_STEM_2.pdf

National Research Council (NRC). (2013). *Monitoring progress toward successful K–12 STEM education: A nation advancing?* Retrieved from http://k12accountability.org/resources/STEM-Education/NRC_Monitoring_STEM_Progress_2013.pdf.

Next Generation Science Standards (NGSS). (2013). *Appendix H—Understanding the scientific enterprise: The nature of science in the next generation science standards.* Retrieved from http://www.nextgenscience.org/sites/default/files/Appendix%20H%20-%20The%20Nature%20of%20Science%20in%20the%20Next%20Generation%20Science%20Standards%204.15.13.pdf

Peters-Burton, E. E., Lynch, S. J., Behrend, T. S., & Means, B. B. (2014). Inclusive STEM high school design: 10 critical components. *Theory Into Practice, 53*(1), 64–71.

President's Council of Advisor's on Science and Technology (PCAST). (2010, September). *Prepare and inspire: K–12 education in science, technology, engineering, and math (STEM) for America's future.* Retrieved from http://www.whitehouse.gov/sites/default/files/microsites/ostp/pcast-stem-ed-final.pdf

Reilly, T. S. (2009). Talent, purpose, and goal orientations: case studies of talented adolescents. High Ability Studies, 20(2), 161-172.

Renzulli, J. S., Koehler, J. L., & Fogarty, E. A. (2006). Operation Houndstooth Intervention Theory: Social capital in today's schools. Gifted Child Today, 29(1), 14–24.

Sands, M. M. (2012). The impact of a peer leadership program on high school students' social capital, as measured by co-cognitive factors of the Renzulli Houndstooth Theory. (Unpublished doctoral dissertation). University of Connecticut, Storrs.

Saunders-Stewart, K. S., Gyles, P. D. T., & Shore, B. M. (2012). Student outcomes in inquiry instruction: A literature-derived inventory. *Journal of Advanced Academics, 23,* 5–31.

Scott, C. (2012). An investigation of science, technology, engineering and mathematics (STEM) focused high schools in the U.S. *Journal of STEM Education, 13*(5), 30–39.

Subotnik, R. F., Tai, R. H., Rickoff, R., & Almarode, J. (2010). Specialized public high schools of science, mathematics, and technology and the STEM pipeline: What do we know now and what will we know in 5 years? *Roeper Review, 32,* 7–16. doi:10.1080/02783190903386553

CHAPTER 6

# CONNECTIONS AND CONTRIBUTIONS THROUGH RESEARCH SERVICE LEARNING

**Terry M. Tomasek**
***Elon University***

**Lacey D. Huffling**
***Georgia Southern University***

**Catherine E. Matthews and Heidi B. Carlone**
***University of North Carolina–Greensboro***

## ABSTRACT

Research-based science combined with service learning is a pedagogical strategy designed to engage students in scientific inquiry while addressing a need in the community. This strategy has the potential to increase students' engagement in science as well as foster students' connections to their community. In this chapter, we explore 16 diverse high school students' meanings of a research-based service learning project that focused on documenting local amphibian and reptile populations for their state museum. Three main themes emerged: students felt they were doing science, they were engaged in

*Science and Service Learning*, pages 125–145

meaningful science work, and they made accomplishments and valuable contributions. The project enabled students to make connections to the larger scientific community as well as their geographical community. Findings suggest that research-based service learning increases students' affiliation with science, which might lead to more sustained participation in science, technology, engineering, and mathematics activities.

High school students ($n$ = 16) elected to take a herpetology class in a college access program. Most had never heard of herpetology—the study of reptiles and amphibians. None could identify frog calls, and most are terrified of snakes. None had ever been to an ephemeral pool. The idea of a mark-recapture study was foreign to them. Couple this background with a statement from a state museum herpetology curator who said the county where the students all live is poorly represented in the state's herpetology specimen collection. Now students had a challenge and a mission. Because of a research-based service learning project, the students might alter their perceptions of these animals and make stronger connections to their local communities and the scientific community as well as improve the state's herpetology record for their county.

This chapter describes the out-of-school science learning experiences of a diverse group of high school students involved in a research-based science service learning project. The students were identified as underrepresented in science, technology, engineering, and mathematics (STEM) and in institutions of higher education: ethnically diverse, low income, and with a limited family history of higher education. The partnership was between students in a summer herpetology science enrichment class that was part of a college-access program, the state Museum of Natural Sciences, and the state wildlife department. The project originated from the underreporting of local reptile and amphibian species and the lack of voucher specimens. Such specimens document the presence of these organisms in the county where the high school students lived, which was the same county where the college-access program was offered. A voucher specimen is any specimen, usually but not always a cadaver, that serves as a basis of study and is retained as a reference, preferably in an accessible collection.

With the guidance of the instructor (one of this chapter's authors), students partnered with the reptile and amphibian collections manager at the state Museum of Natural Sciences. The students became scientists who worked in the field, learned about and collected specimens for the museum to document the presence of organisms in their county. This voucher project provided service to the scientific community because it extended the county records of species and enlarged the state's herpetological collection. The service learning framework for this course had unanticipated positive outcomes for the students. As one youth said, "It's actually doing real science. You're not just sitting in class learning about science from a piece of paper, you're actually doing and contributing to the scientific community." Student engagement in the research-based service learning functioned as a gateway to make science more meaningful and accessible.

We highlight the perspectives of this diverse group of youth who engaged in research-based science service learning. Their engagement was surprising and illuminating as we considered how best to encourage them to pursue higher education; to connect with their community; and to see science knowledge, skills, and ways of thinking as reachable. The research questions guiding this study focused on student meanings of the voucher project and connections students made with the geographical community and scientific communities.

Service learning experiences enhance academic engagement in learning while at the same time building a sense of civic responsibility (Elyer, Giles, Stenson, & Grey, 2001; Sitter, 2006). Research-based science combined with service learning is a specific pedagogical strategy (Reynolds & Ahern-Dodson, 2010) designed to engage students in scientific inquiry that also serves a scientific research need in the community. We describe a research-based science service learning project and present ways that students made unique connections to their geographical communities and became more deeply immersed in the community of science.

## A RATIONALE FOR ACTION

According to the U.S. Department of Education (Kena, 2014), since the turn of the century we have seen some growth in the number of students who have completed high school but less growth in the number of students who pursue and attain science, technology, engineering, and mathematics (STEM)-related degrees. There are still participation gaps among races and also among those from different socioeconomic groups. Between 2000 and 2013, the average graduation rate for public high school students increased from 72% to 81%. In 2013, the graduation rate was 85% for white students, 76% for Hispanic students, and 68% for black students. This presents a graduation rate gap of 9 percentage points between whites and Hispanics and 18 percentage points between white and black students. At the same time, the number of school-aged children living in poverty in the United States increased from 15% to 21%, with the highest percentage of child poverty in the south (23%) (Kena et al., 2014). Of dropouts, 81% reported that if they had been provided engaging instruction in high school with real-word learning that made school more relevant, they would have been more likely to finish school (Bridgeland, DiIulio, & Wulsin, 2008).

There are differences between ethnic groups in STEM field retention. In a study of postsecondary students between 1995 and 2001, there were no measureable differences found between the percentages of black, Hispanic and white students entering STEM-related fields. However, white students had a higher STEM bachelor's degree completion rate than did black and Hispanic students. Parents' education levels were related to STEM-degree attainment. The overall STEM-degree completion rates were higher among students whose parents had at least a 4-year college degree than among those whose parents did not complete that level of education. The STEM attrition rates—students changing their majors from STEM related degree programs—were high. For black students there was a 65% attrition rate, for Hispanic students 50%, and for white students 48%. Those whose parents had high school degrees or less or who came from the lowest 25% income level, had an attrition rate just under 60% (Chen & Weko, 2009). Not only are few students entering STEM-related fields, they are leaving those fields in high numbers.

Schools serving mostly low-income students are less likely to use service learning—29%, compared to 36% of other schools—however, lower-income schools tend to see greater positive impact on student achievement and school engagement than do schools serving students from higher income levels when utilizing service learning pedagogies (Kielsmeier, Scales, Roehlkepartain, & Neal, 2004). Research-based science combined with service learning is one way that students can make better connections with their local communities while engaging in more meaningful ways with science. Service learning may be a way to motivate students to finish high school, attend and complete college, and engage in careers about which they are passionate.

## THEORETICAL FRAMEWORK

We view learning from a sociocultural perspective, which includes a number of assumptions. First, learning occurs through participation in a group's practices (Lave & Wenger, 1991; Wenger, 1998). As newcomers to a community of practice gradually become competent in the group's practices, they come to see themselves as legitimate members of the group. Learning science means engaging in and contributing to its practices. The problem, however, is that often science's practices, as enacted by a fully mature community of professional scientists, may seem "too distant and irrelevant to students" (Brickhouse, Lowery, & Shultz, 2000, p. 444). We propose that adding a service learning component to science minimizes gaps between students and science, highlighting science as a tool to help others and the environment, and making science more relevant and meaningful to youth who participated in the herpetology class.

The second assumption we make about learning is, as participants engage in practice, they not only learn the content and practices of science, they also begin to develop worldviews, dispositions, and identities that are aligned with members in that particular community of practice. Wenger (1998) described this aspect of learning as "learning as becoming," where "learning changes who we are and creates personal histories of becoming in the context of our communities" (p. 5). Youths' perceptions of themselves, herpetology, and reptiles and amphibians changed over time as they became more familiar and competent with the animals and tools and discourses of the discipline.

Third, we consider learning to be a process of meaning-making over time (Wenger, 1998). Participants' engagement in shared practices promotes shared meanings of what is valuable, laudable, undesirable, and deviant. Youths' participation promoted shared meanings of what counted as legitimate and "helpful" practices, and this chapter explores some of those shared meanings.

In the following sections, we provide a description of the broader idea of service learning. We show how service learning has been a pedagogy used specifically in science, and we offer a description of a specific type of science service learning called research based science service learning.

## BACKGROUND

### Service Learning

Service learning (SL) is a teaching and learning strategy integrating meaningful community service with instruction and reflection (Checkoway, 1996). In the best cases, students have the opportunity to identify and address issues of public concern. Service learning enriches students' learning experience and teaches civic responsibility. It has the potential to open students' eyes to the needs of the community and even strengthen the community (Hypes, 2009). Students experience real-life connections between their education and everyday issues when they engage in service learning. Henness (2001) found that students who were engaged in service learning developed significantly more favorable relations with adult civic leaders and community organizations. This was most apparent when the experience pertained to high-priority community issues. High-priority community issues refer to those public issues having communitywide implications that have been identified by community officials or citizen leaders. Learning while serving should transform both the student and the community; however, there is often a struggle to balance these.

Service learning is most powerful when students have the opportunity to develop and plan out their own ideas (Long, 2007). For maximum benefit toward the education goals for students, the community service component of a service learning project should be tightly integrated into the subject matter of the course, understanding that sometimes the "content" may be a process (e.g., participation in authentic scientific practices). Successful projects are also those that maintain room for students to be creative and spontaneous in the learning experience.

Service learning typically differs from volunteerism, internships, or practicums in that during service learning students intentionally use their intellectual capacities and skills to address community problems. While students are gaining knowledge and developing skills, they are, at the same time, using that knowledge and skillset to meet the need within a community. Through reflection on the experience, students come to understand the underlying social, political, environmental, or economic issues that might be contributing to the community challenge (Cress, Collier, & Reitenauer, 2013).

## STEM, Science, and Service Learning

While service learning can be utilized in any subject area, it can have powerful connections to science teaching and learning with the same goals of improving academic engagement and civic responsibility. It provides opportunities for students and adults to collaborate as partners to improve their communities. Involvement in service learning that is science focused allows students to practice dimensions of science literacy (Reynolds & Ahern-Dodson, 2010). As students are building new science knowledge, they can use that knowledge to make decisions about how to help others (McDonald & Kromer, 2005). As students learn about healthy life choices, for example, they may want to communicate this information to others by setting up a booth at a community fair. Students retain scientific information better and understand the complexities of science and society when engaged in a service learning project. Contextualizing science in practical problems that may lead to benefits to society has been shown to spark and sustain a long-term interest in STEM subjects for women (Carlone & Johnson, 2007; Halpern et al, 2007; Knight, Mappen, & Knight, 2011).

There are many examples of science service learning projects. There are beautification projects such as cleaning up city parks, riverbanks, or coastal areas (Schneller, 2008), planting trees, or planting gardens. There are education projects like teaching about nutrition to younger students or helping to find food for monarch butterflies in urban areas (Edwards,

2003). Students may plant a community garden to raise food for the impoverished or organize and participate in recycling efforts (Edwards, 2003; Long, 2007). Students are often highly engaged in these types of projects and have their eyes opened to major issues in a particular field of science. Through these experiences, students see the opportunity to make a difference in the world around them.

One example of science service learning is star parties. Long (2007) described service learning star parties with high school students at a northwest American high school. Students set up telescopes and invited community members to attend information sessions prior to viewing the sky through telescopes. Positive outcomes of the project were increased pride in the students toward their community as well as empowerment of the community to build on the momentum created by students. As students taught community members about the solar system, they were building strong ties to the community through their schoolwork. Students engaged in this project because they wanted to make a difference in their economically struggling community. The star parties were used as a potential tourist attraction to share with those who were visiting in the community.

## Intersections Between Service Learning and Place-Based Education

Place-based education is grounded in the resources, issues, and values of the local community. The focus is on using the local community as a context for learning (Powers, 2004). The term *place-based education* is often used interchangeably with service learning. In both cases, there are intentional connections made between the learner at school and the learner in the community. As with service learning, place-based education helps students see the relevance of what they are learning, thus contributing to their intrinsic motivation and engagement in the learning process (Powers, 2004). Teachers indicate that service learning students are more attentive to their schoolwork and try harder because they know the community is involved and cares about the work they are doing. Service learning students indicate that they learn better when schoolwork has a broader purpose and they are seen by teachers to take on more responsibility and demonstrate higher levels of maturity. A program evaluation conducted by researchers at the Harvard Graduate School of Education for the Rural Trust (1999a, 1999b) found multiple benefits to place-based education: students' academic achievement improves, their interest in the community is increased, and community members are more connected to schools and students.

After evaluating several place-based programs, researchers associated with the Place-based Education Evaluation Collaborative developed a working theory of change for place-based programs (Powers, 2004). The theory suggests that when a person has developed an attachment to their place and then develops the skills to proceed, the individual will become a more active participant in his or her community. This type of civic engagement can nurture a web of relationships that connect individual to community, much like service learning.

## Research-Based Service Learning

The broader idea of service learning is a form of experiential education in which students engage in activities related to needed education outcomes and identified community challenges. Structured opportunities and reflection help students learn from the experiences. Science service learning, as described above, is when students engage in service learning relating to science content. Research service learning (RSL) refines the model of science service learning by including an element of scientific research. As an emerging pedagogy, RSL engages students in authentic, community-based scientific research while showing students how to use their new knowledge to serve society (Reynolds & Ahern-Dodson, 2010). Students learn to ask research questions relevant to their communities' needs and then work with teachers and community partners to design and implement research projects to address identified needs. Partnering with a local state park to conduct a small mammal investigation or map the distribution of invasive plant species are possible RSL projects. The park identifies the need for this type of scientific data and students provide the information by conducting the needed investigation. Research service learning helps students understand the demands of scientific research, cultivates critical-thinking skills, and increases understanding of environmental issues (Reynolds & Ahern-Dodson, 2010). Students benefit by studying authentic scientific questions, and communities benefit by receiving answers to pressing environmental questions. Students are treated as colleagues, not just volunteers, while faculty are not just teachers, but collaborators and co-learners.

The service learning work presented in this chapter is a type of research service learning. Students partnered with the state museum to conduct a biological inventory in the county where they lived. This type of inquiry involves making careful observations and reporting accurate information. A biological inventory begins by developing a simple list of species. The inventory expands as information about the ecological community is included. An inventory is normally taken over a short period of time and represents a snapshot of a population or habitat. A good inventory can produce baseline data, which is the standard against which comparisons can be made over time.

## DESCRIPTION OF THIS RESEARCH SERVICE LEARNING PROJECT

The service learning project framed the herpetology (the study of amphibians and reptiles) course provided to high school students who were part of a college-access program. Students participating in the college access program were underrepresented on college campuses; they were low-income students, ethnically diverse, or had limited family college experience. Students took part in a 4-week residential experience on a university campus. During this program, students selected academic courses, one choice of which was a herpetology course. During the summer of 2012 there were 16 students in the course, nine females and seven males. They were high school sophomores, juniors, and seniors. Native American comprisesd 6%, 25% were Hispanic/Latino, 31% were Caucasian, and 38% were African American. This demographic range is similar to the Academy as a whole ($N = 97$). All students came from a single rural county in the southeastern United States; the same county where the college that conducts the college-access program is located.

The herpetology course met four days per week for four weeks, two hours per day. Three days per week were classroom activities while the fourth day each week was a field science investigation. The purpose of the course was to introduce students to local reptile and amphibian species, engage them in several ongoing scientific inquiry studies, and provide them with a foundational knowledge of the natural history of amphibians and reptiles, including global and local environmental issues. Students developed skills in critically reading and reviewing popular media, reflective scientific journals, news resources, and published research. Students gained valuable research skills in collecting, reporting and interpreting scientific data, taking field notes, identifying organisms and evaluating habitat. They were exposed to careers in the biological/ecological field science disciplines and met and worked with a number of herpetologists. A major focus of the course was to help students develop an awareness of common amphibians and reptiles in the local environment.

The idea of the service learning project began when the course instructor was attending a statewide professional meeting. One of the speakers talked about the value of voucher specimen collections to provide a permanent record of species diversity in the state. A voucher specimen is a preserved representative from a taxonomic group that serves as a future reference for scientific inquiry. The instructor contacted the collections manager for herpetology at the state Museum of Natural Science to talk about a possible collaborative project. The collections manager indicated the students' county of residence was indeed one of the more poorly documented counties in the state. He indicated that there was value in partnering with the students to contribute to the museum collection and he called this a high-priority issue. Even though citizens of the county may not have been aware of the need, the community, defined as the community of scientists, indicated that this was a high-priority need.

At the beginning of the course, students learned about general characteristics of these animals and habitat preferences based on these characteristics. For example, salamanders (amphibians) have smooth moist skin that allows for exchange of gases. This means that most salamanders will be found in moist environments. Students had the opportunity to observe live animals in the classroom to reinforce their learning. During one class session, each table group ($n = 4$) had a container in the center of the table with a species of frog. Students learned about morphological features by observing and sometimes holding the animals. Although the focus of the course was to introduce students to local species, there was also an emphasis on global issues that effect populations of animals worldwide. Habitat degradation, for example, impacts an area's ability to sustain a population of animals. Habitat degradation was examined at the local level as well as nationally and internationally. Climate, on a global scale, is correlated to some amphibian declines and may also be related to changes in local habitats (i.e., certain temporary pools drying up quicker than is typical).

Students were given a field guide and asked to skim through the different groups of animals and make a list of those animals that should be present in the county based on range maps. After doing this, students used county-level dot-distribution maps to see where frogs had actually been recorded across the state. They did the same for other types of amphibians and reptiles by accessing the statewide citizen science database to compare their list of organisms to those reported. Students found very few records of the animals that should have been in the county. A class discussion was used to talk about the meaning of these apparent conflicting findings. Students were quick to wonder about the validity of the field guide, believing initially that information found online was the most accurate. As the discussion continued, students started to develop possible explanations for why there was a great deal of inconsistency between the online database and their list generated from the field guide.

At this point, the instructor shared with students an email written by the collections manager for herpetology from the state museum. This person was also one of the authors for their field guide so students had an immediate connection. The email began with a statement that the county of residence for these students was indeed "one of the many poorly documented" counties in the state. He also suggested that a reason for this might be that neighboring counties (highly populated) have had most of the good habitat for reptiles and amphibians destroyed, which could possibly impact populations in the students' county. The county contained virtually no public lands or large chunks of pristine habitat. He provided a list of species that the museum's records lacked from the county and indicated that he would be "glad to get specimens of any of those species." He also indicated that photographs or reliable observational records would be useful. In addition, he provided helpful suggestions for the best way to transfer live specimens to the museum and how to preserve salvaged specimens.

Students were appalled that their county was considered poorly documented; however, they had seen that for themselves in the dot-distribution maps and the online database. They were quick to want to be part of the solution. The class brainstormed a plan for developing this partnership; they immediately recognized that they needed to learn more about local species of amphibians and reptiles. During the remaining 4 weeks of the program, the instructor systematically began to teach students about the local diversity of animals and the relationship between habitat and biodiversity. Two students in the class indicated that they wanted to take on the leadership role of the project, which included keeping our partner, the collections manager, informed about our progress. They kept a written record of each type of animal needed as well as when and how many specimens were delivered to the museum. In addition to learning about biodiversity and habitat use, students also engaged in a variety of scientific inquiries related to population studies, such as calling amphibian surveys to better understand local populations of frogs and toads; ephemeral pool surveys to better understand breeding patterns of salamanders and frogs; aquatic turtle trapping involving a mark/recapture study in a local lake; and checking coverboard transects for snakes, lizards, and other forest amphibians. Descriptions of these various projects have been published elsewhere (Matthews & Tomasek, 2012a ,2012b; Tomasek, Matthews, & Hall, 2005; Trautmann, Fee, Tomasek, & Bergey, 2013).

The class brainstormed places in the county where there was good habitat for the types of animals for which they were looking. The instructor provided a county map, and students placed pins to indicate where they lived. Students talked about local neighborhoods describing where in their neighborhood they might find animals. Over the 4 weeks of the program, the class collected needed animals for the voucher specimen project. Each Saturday of the program, all students left campus to go home for the evening and returned to campus the next afternoon (Sunday). This break gave students the opportunity to explore their own neighborhoods for needed organisms. Many of the students returned to class each Monday bringing animals in plastic containers. Students with animals shared how they found, captured, and identified the animals. The class confirmed the identification of the animals, and the two student leaders made preparations for transporting animals to the museum.

At the end of the summer program, the class traveled to the museum to present their voucher specimens to the collections manager. Students asked the collection employees knowledgeable questions and interacted with the collection manager in competent and confident ways. For example, students quizzed each other on animal identification before reading labels. They also made careful observations of the growing collection, noting animals that they recognized and those they did not. For well over an hour, the collection manager told the students "stories" about how he had found many different types of animals. In return, the students shared their "stories" with the collections manager.

## DESCRIPTION OF RESEARCH STUDY

The research questions guiding this study were "What were student meanings of the voucher project?" and "What connections do students make with their geographical community and with the scientific community?" By geographical community we mean the areas where students lived, attended school, and engaged in out-of-school learning experiences. We argue that student participation in this research service learning project provided them the opportunity to make connections with their local community as well as with the scientific community. These types of connections may help students to see themselves as civic and scientific individuals in new ways.

We collected data directly from students via surveys and interviews. At the end of the herpetology course, we conducted 45-minute interviews with each of the 16 participants. During each interview we implemented a card-sort activity (see Carlone, 2012; Carlone, Haun-Frank, & Webb, 2011), which prompted participants to rank and describe their experiences with some of the course's cultural norms and practices. We asked them to decide whether or not the norm or practice on each card was an important part of being a good participant in the course, to describe the meaning of that norm or practice as it related to their experiences in the course, and then to provide us with an example or two of the norm or practice. This allowed us to understand their meanings and valuations of each norm and practice. The cards that were helpful in our analysis included make careful observations, collect data, help others, think like scientists, ask questions, discuss science as a tool to help people or the environment, make mistakes, use scientific vocabulary, use scientific tools, be precise or accurate, talk like scientists, and be curious. We also asked students about their perceptions of and learning afforded by their participation in the research service learning project. All interviews were audiotaped and transcribed. We gave students a written pre/postsurvey with Likert-scale statements aimed at understanding the ways they experienced, understood, and valued the various activities in the herpetology class and their science interests and affiliations.

## RESEARCH FINDINGS AND DISCUSSION

When students described their meanings associated with the research service learning voucher project, they focused on three main themes: doing science, meaningful science work and feelings of accomplishment, and value of contribution. First, students saw the project as an opportunity for them to find missing herps—reptiles and amphibians—in their county (doing science). As a result, they understood their responsibility and ability to contribute to museum records for the county where they lived (meaningful science work). Finally, students were excited and proud of their contributions describing how it made them feel like a scientist (feeling of accomplishment and value of contribution).

All participants were aware that the voucher project involved finding and collecting specimens that were not already documented in their county of residence. In a sense, students embraced the challenge of finding that which appeared to be hidden. There was an element of adventure but unlike the typical scavenger hunt, students felt they were making "important finds," "contributions to science," and even "helping the scientific community." These feelings provided intrinsic motivation and a greater attention to detail than might have been typical. "We are actually out in the public, finding out something nobody knew was here" (Katie interview).

"I really like the fact that I am actually contributing to science with it [the voucher project]" (Beatrice interview). "I felt like we were helping, especially... the different herpetology researchers... because they say that Adams County [a pseudonym] was... the lowest recorded herp thing in the state" (Chris interview). Student responses on the end-of-program survey were also extremely high and related to how they came to see themselves as people who could contribute to scientific data. The average ratings for statements related to comparisons of self and science—1 (not at all) to 5 (to a great extent)—are reflected in Table 6.1.

**TABLE 6.1 Comparisons of Self and Science**

| | |
|---|---|
| The learning experience made me feel like a science person | 4.56 |
| I can contribute to science | 4.60 |
| I can use what I have learned about science outside of school | 4.88 |
| I can think like a science person | 4.63 |

In multiple instances, youth described their role as helping out the scientific community. In this manner, the service learning work was meaningful. They saw a need in the scientific community and they were coming alongside that community to offer assistance. Students understood that the knowledge they gained (and were sharing) had a bigger purpose.

> I thought it was really cool that we had the chance as high school students to help out such a larger project and to make a contribution. It was really interesting to know that we've seen these species but no one's recorded them before now and for us to be able to do that... that's just mind blowing... as high school students you don't think that you can, like, you think that's a scientist's job that you have to go to college for. (Serenity interview)

"I felt accomplished that we were one step closer to helping out the scientific community in getting those specimens that are supposed to be found in Adams County actually recorded" (Carson interview). "You're helping a bigger purpose and you know everybody's important, and it makes you feel good" (Kalyn interview). "I'm pretty sure that [collecting data] was the most important thing that we were going to do in this class, since we're helping the state do things and maybe the nation" (Jose interview).

Others' perspectives on the project highlighted how the voucher project brought them recognition collectively. The youth felt like their contributions were important not only to science but to the community in which they lived. Students believed that their work was making a difference.

> Actually it [the voucher project] meant a lot to me because it's like I could actually do something in class that would affect an actual bigger scale and we can actually get Adams county on the map, so to speak, that we'd have these specimens recorded in Adams county and it's actually doing real science, you're not just sitting in class learning about science from a piece of paper, you're actually doing and contributing to the scientific community, even though it's a small part, it's a part. (Carson interview)

"It actually felt good for me to actually make a difference and put their [reptiles and amphibians'] name on the map" (Theo interview). "We actually made a difference. We are in the records of North Carolina. Cause that's just something that you don't ever do" (Gabriela interview).

Students' efforts gave them a sense of accomplishment and pride in making sure that the county where they lived was better represented in the museum collection. Eva defined the voucher project like this: "In here [the herpetology course], we go out—outside—and do fieldwork, and we actually collect things, and then we go to give the information to somebody who's gonna put them to a bigger use." The students experienced research service learning in that they recognized that they were being trusted to engage in and carry out scientific research in the form of a biological inventory. "We're actually contributing to what's going on around us. I felt really involved in what's going on. It's not we're just studying reptiles and amphibians; we're actually being part of what's going on, of the data that's being collected" (Maria interview).

Students made stronger connections with their county's natural areas and wildlife than they had before the course. They developed a deeper appreciation for the native reptiles and amphibians (herpetofauna), often expressing surprise and amazement at the abundance and diversity of their county's herpetofauna of which they were previously unaware. The voucher project nurtured students' motivation to practice and educate others about good environmental stewardship to protect habitats for herpetofauna that they encountered in their fieldwork.

> I didn't know so many little creatures existed out there until I took this herpetology class and now when I'm out, or at night, I'm actually attentive to hear and see and put my learning skills that I've learned to practice at home or wherever I go and then being able to share it with some of my younger siblings or anybody just makes me really proud of myself to know that knowledge and to be able to share it with the people I love. (Valarie interview)

They were delighted about their new level of knowledge and skill in identification and began to see their worlds in new ways. "And now I know that there are frogs and turtles and everything like that hopping around everywhere" (Gabriela interview). "I never knew how to tell the difference in the sex of a turtle, for example, or the differences in snakes and frog sounds because I hear frogs all the time in my house and I could never pick out which one's what and now when I go home on the weekends I can go, oh that's a cricket frog, oh that's a fowlers toad" (Carson interview). "I was with my friends and it was really dark and I was like 'gasp' that's a frog. That's a northern cricket [frog]" (Gabriela interview).

Their newfound knowledge of biodiversity in their home communities cultivated a deeper appreciation for that diversity. With respect to local biodiversity, the students' world expanded. What once was silent or simple had now become noisy and complex, in a good way. As a result of their service learning experience, students now lived in a richer, more diverse world. Students had not only become more aware of their natural surroundings, but they had become discerning of the intricate nature of their surroundings, gaining affinity for the local herpetofauna and developing stronger ties to their community.

> I would have never known that there was a frog in [the local] park. I thought that frogs were all neon green. I didn't know that there was just so many there and so many that we didn't know about in Adams County. (Beatrice interview)

> Before, I would never listen to it [frog calls]. I would just listen to the cars passing by, that's all. But then I block off that car sound and I would hear all these animals, the frog calls, and there's different kinds of frogs. It's really cool. (Jamie interview)

> I memorized the sound now and I can go out at night and be like, oh, that's a Fowlers toad and that's a Northern cricket frog and I think just sitting out at night and listening to them is peaceful and there's just something about frogs that I just fell in love with. I definitely love them. (Kalyn interview)

In some instances, their awareness of the local environment coupled with a reduction of fears about herps, was growing into concern for animals and the natural world, which has led to the beginnings of environmental stewardship. Participants were determined that their county needed to practice good environmental stewardship so these animals continue to have a place to live.

In addition to the interview data, student surveys indicated a greater connection to nature, greater interest in nature, and more empathy for animals as well as a greater understanding of threats that reptiles and amphibians face. The average high ratings for statements related to connections to nature—1 (not at all) to 5 (to a great extent)—are reflected in Table 6.2. Students' eyes were opened to their community's needs, including the community of amphibians and reptiles.

> As I'm walking, even as we're walking to lunch or stuff, I'll look around at the plants and stuff, and that's something I would have never done. I'm not as scared as I used to be because I actually know what's going on with it. I'm not in danger by these little creatures. (Gabriela interview)

"It was important to me to realize that I could be doing something in my own home to help out the general population of amphibians" (Carson interview). "[We should study reptiles and amphibians] because they matter... Fear is associated with them but they are still important to the ecosystem... If something happens to them, then we are in trouble" (Chris interview). Students acknowledged that what they were learning was directly related to their community. Students were proud of their community and wanted their county to be an active participant in the work of doing science. They reveled in their discoveries and were anxious to share their

**TABLE 6.2 Connections to Nature**

| *To what degree did participating in this herpetology research experience increase your:* | |
|---|---|
| Interest in nature? | 4.94 |
| Connection to nature? | 4.5 |
| Empathy for animals? | 4.67 |

discoveries with others and to even continue the work outside of the group. They were able to do this because their knowledge of the local environment had been expanded.

> I know more about what's out there now. So there's a lot of new places that I know where things are at and now I know where to go if I wanted to try and catch them and put them down and help the [state database]. So now I know where to go and what to do. (Kalyn interview)

The theory of change for place-based programs described by Powers (2004) indicates that an attachment for place combined with skill development can lead to more active participation in the community. This research-based service learning project provided students the opportunity to enhance their sense of place. Participants also developed skills and knowledge related to local species diversity. A result was active participation in their geographical community as well as the community of scientists.

As indicated by the students' meanings and perspectives of the voucher project, students readily made connections between the voucher project and the larger scientific community. The participants connected to the scientific community by (a) helping to collect voucher specimens for their county, which enabled them to make contributions to science; (b) viewing themselves as potential scientists; and (c) engaging in scientific practices, students identified with the larger scientific community.

As Serenity discussed above, "you think that's a scientist's' job" and Eva said that the voucher project was "not just learning about stuff, but actually being able to contribute." Continuing her description, Eva stated, "In here [herpetology class], we do fieldwork, and we actually collect things, and then we go to give the information to somebody who's gonna put them [voucher specimens] to bigger use." The participants also made personal connections with the scientific community such as when Carson described himself as feeling "accomplished that we were one step closer to helping out the scientific community; and Beatrice also indicated that her enjoyment of the project stemmed from "the fact that I am actually contributing to science with it [voucher project]."

As discussed in our previous research questions, the students' responses on the end of the program survey were extremely high for how the class made them feel like a science person and how it was possible to contribute to science, use what they know about science outside of school, and think like a scientist (see Table 6.1). Ezra stated in his end of program interview, "Technically, I count us as professionals because we're taking data and observing the animals, capturing animals, and we're doing a lot of things."

Participants also connected specific scientific practices with the contributions to the science they were making through the voucher project.

Making careful observations, collecting data, and being precise or accurate were the three scientific practices participants most often discussed in relation to the voucher project. Participants discussed how making careful observations were important because they had to identify the organism they caught and know that their identification was correct. As Gabriela observed,

> I would say we were expected to do that [make careful observations] because when we were finding animals or our voucher specimens, we had to make sure that they were the right species because some of them looked similar and we didn't want to get them confused.

The participants realized that making careful observations was required for their contributions to science to be recognized as valid by other scientists. Collecting data was often discussed in conjunction with making careful observations, as participants often viewed these two practices as linked. As shared earlier, Jose linked collecting data with helping the state, and Chris stated that the participants had to collect data to "get Adams County on the map." Participants viewed being precise and accurate as essential to the success of the voucher project. As Cassie shared, "The information has to be correct so people won't go back and say 'Well, you're not really certain about this so why should we trust you, why should we believe you?'"

Though these three practices are sometimes seen in classroom scientific practices, the participants more readily connected these practices with the contributions to the science they were making with the voucher project. Thus, these shared practices allowed participants to connect to the larger scientific community and afforded the participants opportunities to engage in research service learning. In addition to the interview data, student surveys indicated a greater interest in and confidence in doing science, a greater interest in participating in science, and an increased ability to use scientific tools. The average ratings for statements related to connections to science—1 (not at all) to 5 (to a great extent)—are reflected in Table 6.3.

**TABLE 6.3 Connections to Science**

| *To what degree did participating in this herpetology research experience increase your:* | |
|---|---|
| Knowledge of science? | 4.50 |
| Confidence in doing science? | 4.63 |
| Interest in science? | 4.94 |
| Ability to use scientific tools? | 4.50 |

## IMPLICATIONS

The voucher project gave underrepresented students the opportunity to engage in authentic, community-based scientific research. Students acquired knowledge and skill at the same time that they were using that knowledge and skill to serve the scientific community and learn about their local community. Contextualizing science in the needs of the community helped this group of diverse students develop a greater interest in science, which may lead them to want to pursue a STEM-related career. This is important to the discipline of science because economically disadvantaged youth deal with different types of problems and can therefore bring different perspectives to problem solving that is at the heart of scientific inquiry. Research-based service learning may be a form of culturally relevant pedagogy for underrepresented students in STEM. Engaging in this research-based service learning project opened students' eyes to the needs of the community and provided a pathway of entry into the scientific community, creating a bridge between the world of student and scientist. The service learning experience gave students an authentic science connection to their everyday lived environments in a way that increased their understanding of environmental issues including biodiversity. Students took pride in their scientific accomplishments, while at the same time they came to see the world differently.

## ACKNOWLEDGMENTS

This work was supported by the National Science Foundation under Grant ISE#1114558. Any opinions, findings, and conclusions or recommendations expressed in this manuscript are those of the authors and do not necessarily reflect the views of the National Science Foundation.

## REFERENCES

Brickhouse, N. W., Lowery, P., & Schultz, K. (2000). What kind of a girl does science? The construction of school science identities. *Journal of Research in Science Teaching, 37*(5), 441–458.

Bridgeland, J. M., DiIulio J. J., Jr., & Wulsin, S. C. (2008). *Engaged for success: Service learning as a tool for high school dropout prevention.* Washington, DC: Civic Enterprises.

Carlone, H. B. (2012). Methodological considerations for studying identities in school science: An anthropological approach. In M. Varelas (Ed.), *Identity construction and science education research: Learning, teaching, and being in multiple contexts* (pp. 9–26). Rotterdam, The Netherlands: Sense.

Carlone, H., Haun-Frank, J., & Webb, A. W. (2011). Assessing equity beyond knowledge- and skills-based outcomes: A comparative ethnography of two fourth-grade reform-based science classrooms. *Journal of Research in Science Teaching, 48*(5), 459–485.

Carlone, H. B., & Johnson, A. (2007). Understanding the science enterprises of successful women of color: Science identity as an analytic lens. *Journal of Research in Science Training, 44*(8), 1187–1218.

Checkoway, B. (1996) Combining service and learning on campus and in the community. *Phi Delta Kappan, 77*(9), 600–606.

Chen, X., & Weko, T. (2009, July). Students who study science, technology, engineering, and mathematics (STEM) in postsecondary education. *NCES.* Retrieved from http://nces.ed.gov/pubs2009/2009161.pdf

Cress, C. M., Collier, P. J., & Reitenauer, V. L. (2013). *Learning through serving: A student guidebook for service learning and civic engagement across the disciplines and cultural communities.* Sterling, VA: Stylus.

Edwards, J. (2003). Butterfly business: Connecting science and service learning. *Investigating, 19*(1), 8–11.

Eyler, J. S., Giles, D. E., Jr., Stenson, C. M., & Gray, C. J. (2001, August 31). *At a glance: What we know about the effects of service-learning on college students, faculty, institutions and communities, 1993–2000: Third edition.* Retrieved from http://www.compact.org/wp-content/uploads/resources/downloads/aag.pdf

Halpern, D. F., Aronson, J., Reimer, N., Simpkins, S., Star, J. R., & Wentzel, K. (2007, September). Encouraging girls in math and science. *Institute of Education Sciences.* Retrieved from https://ies.ed.gov/ncee/wwc/pdf/practice_guides/20072003.pdf

Harvard Graduate School of Education for the Rural Trust. (1999a). *Living and learning in rural schools and communities:* A report to the Annenberg Rural Challenge. Cambridge, MA: Harvard Graduate School of Education.

Harvard Graduate School of Education for the Rural Trust. (1999b). *Living and learning in rural schools and communities: Lessons from the field.* Cambridge, MA: Harvard Graduate School of Education.

Henness, S. (2001, July). K–12 service learning: A strategy for rural community renewal and revitalization. *Corporation for National Service.* Retrieved June 2, 2014, from https://4-h.org/wp-content/uploads/2016/02/Service-Learning.pdf

Hypes, M. G. (November 2009). What's in it for me? *Journal of Physical Education, Recreation and Dance, 80*(9), 6–7.

Kena, G., Aud, S., Johnson, F., Wang, X., Jijun, Z., Rathburn, A., . . . Kristapovich, P. (2014, May). The condition of education 2014. *NCES.* Retrieved July 21, 2014, from http://nces.ed.gov/pubs2014/2014083.pdf

Kielsmeier, J. C., Scales, P. C., Roehlkepartain, E. C., & Neal, M. (2004). Growing to Greatness 2004: Community service and service learning in public schools. *National Youth Leadership Council.* Retrieved from http://www.peecworks.org/peec/peec_research/01795BFB-001D0211.3/growing%20to%20greatness%202004.pdf

Knight, D. B., Mappen, E. F., & Knight, S. L. (2011). A review of the literature on increasing the representation of women undergraduates in STEM disciplines through civic engagement pedagogies. *Science Education and Civic Engagement, 3*(1), 36–47.

Lave, J., & Wenger, E. (1991). *Situated learning: Legitimate peripheral participation.* New York, NY: Cambridge University Press.

Long, C. (2007, November). Making the connection: Service learning links curriculum to the real world. *NEA.* Retrieved from http://www.nea.org/home/10480.htm

Matthews, C. E., & Tomasek, T. (2012a). Professional development and authentic science education: Monitoring local populations of reptiles and amphibians. In R. E. Yager (Ed.), *Exemplary science: Best practices in professional development.* Arlington, VA: NSTA Press.

Matthews, C. E., & Tomasek, T. (2012b). Herpetology for high school students. *Green Teacher, 96,* 36–40.

McDonald, J., & Kromer, T. (2005). Service learning: A way to connect science to the community. *Science Scope, 28*(7), 46–48.

Powers, A. L. (2004). An evaluation of four place-based education programs. *Journal of Environmental Education, 35*(4), 17–32.

Reynolds, J. A., & Ahern-Dodson, J. (2010). Promoting science literacy through research service learning: An emerging pedagogy with significant benefits for students, faculty, universities and communities. *Journal of College Science Teaching, 39*(6), 24–29.

Schneller, A. J. (2008). Environmental service learning: Outcomes of innovative pedagogy in Baja California Sur, Mexico. *Environmental Education Research, 14*(3), 291–307.

Sitter, C. L. (2006). Learning by serving. *Knowledge Quest, 34*(5), 23–26.

Tomasek, T., Matthews, C., & Hall, J. (2005). What's slithering around on your school grounds: Transforming student awareness of reptile and amphibian diversity. *American Biology Teacher, 67*(7), 419–426.

Trautmann, N. M., Fee, J., Tomasek, T. M., & Bergey, N. R. (2013). *Be a scientist! Bring biology to life through citizen science.* Arlington, VA: NSTA Press.

Wenger, E. (1998). *Communities of practice: Learning, meaning, and identity.* New York, NY: Cambridge University Press.

CHAPTER 7

# SOILS IN THE CITY

**Christie Klimas, James Montgomery, Howard Rosing, Marji Hess, Xochyl Perez, and Christian DeKnock**
***DePaul University***

## ABSTRACT

This case study on *Discover DePaul* represents an initial step in responding to the question, How can we use science and education to help revitalize low-income communities, involve and mentor underrepresented students in scientific research in their communities, equip undergraduates with the skills to work with and learn from community partners, and use results to help inform environmental policy at the municipal level? The project was implemented as a 3-week summer science immersion for high school students with a goal of assessing urban soil quality in underserved communities, using soil quality assessment as a portal into scientific exploration by youth in these communities. The itinerary balanced science learning, use of scientific knowledge to answer local questions of interest, and an introductory college experience. Postsurvey results highlight multiple metrics of success with the majority of students self-reporting higher college-readiness and a gain or improvement in 18 of 20 employability skills.

Community based service learning, participatory action research (PAR), and the scholarship of engagement are becoming commonplace in the academy. The literature suggests that scholars have begun seeking a deeper

*Science and Service Learning*, pages 147–178

and more critical understanding of how community engagement in higher education impacts both students and their communities served (Butin, 2015; Hoy & Johnson, 2013; Mitchell, 2015; Reed & Butler, 2015; Stoecker & Tryon, 2009). We report on a community-based soil project conducted in a low-income, predominantly African American Chicago neighborhood with significant vacant lots. The project was implemented collaboratively with the Gary Comer Youth Center (GCYC) located in Chicago's Grand Crossing neighborhood on the city's south side. As means to contextualize the project, the following paragraphs outline the salient seminal literature, including pertinent case studies that have been influential in the pragma and praxis of our community-engaged research work.

A long history of, and scholarship on, university engagement with neighborhoods such as Greater Grand Crossing in Chicago. American educators and philosophers William Rainey Harper and John Dewey informed such engagement, advocating for a praxis education model as a means to promote participatory democracy in the context of 19th century Chicago's social, environmental, and cultural milieu. Harper placed great emphasis on his university's active engagement with the problems confronting the rapidly growing and industrializing city of Chicago. He started the Department of Pedagogy at the University of Chicago for the express purpose of having faculty actively engage with the Chicago public school system to insure that students were given a holistic education geared toward the promotion of democracy (Westbrook, 1991). Dewey's student Robert Park developed a foundation for the study of society that became the theoretical basis for the Chicago School of Sociology. University of Chicago researchers conducted urban and community studies on poverty and immigrants, among other issues. The survey and mapping methods and, more importantly, the way communities and groups were conceptualized and framed as objects of study were groundbreaking for the field of sociology. Moving beyond the study of community, however, Dewey's worldview encouraged a form of education that led to societal transformation through an engaged pedagogy that embraced the concept of "participatory democracy." Dewey laid out the foundations of a new social theory of a participatory democracy in his 1888 essay "The Ethics of Democracy." He also recognized that the achievement of desirable societal outcomes required the integration of academic theories and practical reason; the unity of scientific theory and practice, and the unity of intelligence and action (Benson, Harkavy, & Puckett, 2007).

In particular, Dewey's philosophical and theoretical views on the role that schools should play in the creation of a participatory democracy were radically altered as a result of his strong relationship with Chicago social reformer and settlement house pioneer Jane Addams. Addams and other settlement workers expanded the reach of social, cultural, and recreational services from settlement houses into the schools, realizing that the larger number of schools could reach many more people than the scant few settlement houses. Dewey became enamored with the idea of the school acting as a "social center." Together, he and Addams focused their efforts on improving the conditions and quality of the natural and social environment in the impoverished neighborhoods in Chicago, for they recognized that the creation of a high-quality school system relies on stable and sustainable natural and social environments (Benson et al., 2007). We highlight the pragma and praxis of Harper, Dewey, and Addams because even today, over 100 years after their work, many of the same problems they confronted (e.g., violence, poverty, underachieving schools, out-migration, and general disinvestment) continue to impact Chicago's neighborhoods. The location of our project, Greater Grand Crossing, lies in the heart of such neighborhoods where schools remain as one of the few anchor institutions.

Following Dewey and Harper, later advocates of organizational and community change via action research included scholars such Kurt Lewin and Eric Trist. Lewin was a German social psychologist who is considered the father of action research (Greenwood & Levin, 1998). Based on his work with the American Jewish Congress on minority issues, he famously declared, "No action without research; no research without action (as quoted in Marrow, 1996, p. 193). Trist, a British psychologist and social ecologist, was profoundly influenced by Lewin's ideas. His ideas and applied research in social and organizational ecology were developed during his time as a clinical psychologist at the Tavistock Institute of Human Relations in Great Britain, where he and his colleagues advocated for social change via experimentation, collaboration, and a long-term systems approach (Pasmore & Khalsa, 1993; Trist, Emery, Murray, & Trist, 1997). Trist anticipated that the transition to post industrialism would bring about changes in social patterns, including the *primacy of collaborative relations* (authors' emphasis) (Emery & Trist, 1973).

Uri Bronfenbrenner (1979) extended the work of Lewin, Trist, and others to develop an ecological approach to the study of human development. His definition of ecological development is particularly foundational to our project as it clearly recognizes that the relationship between the natural and sociocultural environments is key to human development:

> The ecology of human development involves the scientific study of the progressive, mutual accommodation between an active, growing human being and the changing properties of the immediate settings in which the developing person lives, as this process is affected by the relations between these settings, and by the larger contexts in which the settings are embedded. (p. 21)

Bronfenbrenner's ideas undergird the concept of human flourishing, a concept which can be traced back to Aristotle's series of lectures, known as the Nicomachean Ethics, delivered in 4th century Athens.

By far one of the most influential scholars on the development of pedagogy infused with praxis is educational philosopher Paulo Freire; his ideas represented the means by which the purpose of research could be fully understood as a means for empowering communities. In *Pedagogy of the Oppressed* (1970), Freire developed the notion of "conscientisation" (*conscientizacao*), a concept involving development of critical consciousness through the practice of reflection and action (praxis) that leads people to engage in seeking social change for themselves. "Conscientisation" emerges from an educational practice that illuminates social, economic, and political conditions "in a situation" (Freire, 1970). The result is deeper awareness of the nature of inequality and oppression in a given context. This deeper awareness, according to Freire, is the basis for action as part of a cyclical process that involves continued return to critical analysis of the social, economic and political context that in turn, inform further research. Freire laid out the foundation for participatory action research (PAR), which evolved into a variety of pedagogical approaches that have been increasingly adopted by the academy in recent decades (Rosing & Hofman, 2010; Strand, Culforth, Stoecker, Marullo, & Donohue 2003).

Some of the earliest efforts in the United States to apply PAR and interdisciplinary teams to solving social problems to enhance human flourishing are found in the work of Marie Kennedy (1995) and her work with the Roofless Women's Action Research Mobilization project; the Policy Research Action Group organized by Nyden, Figert, Shibley, and Burrows (1997); Stanton, Giles, and Cruz (1999); Furco (2001); the Urban Nutrition Initiative conducted by Johnson, Harkavy, Barg, Gerber, and Rulf (2004); Peters (2005); the West Philadelphia Project conducted by Benson et al. (2007) at the University of Pennsylvania; the Great Cities Institute founded by Perry and Wiewel (2008), and the East St. Louis Action Research Project organized by Reardon (2000).

The work of these university-based scholars addresses, in some ways, issues of human development, organizational behavior, and human flourishing. Disinvestment in many low-income neighborhoods of color, replete with abandoned housing tracts, myriad forms of pollution, crime, and misery, is but one causal factor in the stunting of human development and flourishing. One cannot easily achieve a high quality of life (monetary and otherwise) if they live in a milieu of environmental degradation, poverty, and a lack of basic human necessities. The land, particularly the soil quality in such neighborhoods is a fundamental component of this environmental context. How people can engage with the soil for purposes of both education and community development is a driving focus of our research.

The aforementioned action research approaches, informed by the social justice orientation of PAR, are becoming an integral part of the academy through, among other areas, approaches such as community-based participatory research in the health fields (Asselin & Brisbois, 2015; Buff et al., 2015; Israel, Schulz, Parker, & Becker, 1998; Mi, Stefaniak, & Afonso, 2014). Over the past two decades the service learning literature, itself having roots in the work of Dewey, Lewin, and Freire among others, began to integrate and/or advocate for action research as an approach and pedagogy. As higher education began to recognize forms of community-based research as an important paradigm for promoting social change (Strand et al., 2003), college instructors began to creatively integrate research driven by community partners into curriculum in ways that served both student learning and community interests (Hofman & Rosing, 2006). The scholarship derived from these approaches was influential in the design and execution of our project.

## COMMUNITY-UNIVERSITY PARTNERSHIPS AND SERVICE LEARNING FOR PROBLEM SOLVING IN URBAN AREAS

The largest Catholic university in the United States, DePaul University, is guided by its mission built on the life's work of Saint Vincent De Paul (1581–1660). Vincent de Paul's legacy deeply informs the ethos of the university and establishes the basis for institutional support and ultimately the pragma and praxis of our project. He was a French priest and founder of the Congregation of the Mission (popularly known as the Vincentians). Known within the Roman Catholic Church as the Apostle of Charity, he lived almost all of his adult life in Paris, ministering to the material and spiritual needs of the urban poor. He wanted to do something that would change a world in which suffering, injustice, and poverty were endemic and dismissed. His mission was to provide service to people and communities that was empowering and liberating. He understood that making a difference meant challenging the status quo. He also understood the nature of contemporary problems, and he responded with assistance efforts that were part of long-term solutions. In contemporary terms, he was values-driven. He was connected to the community because he knew that community transformation demanded community involvement, a major tenet of PAR and social ecology. Vincent was supportive of diversity and constantly concerned about including those who found themselves excluded by society, by the economy, by the church, or by any human system. A risk taker, he was innovative, and he was pragmatic. He practiced sustainability at a time when this word was not present in the French or English lexicon. As a consequence of Vincent de Paul's work, DePaul University is the kind of institution that puts its mission into practice through ensuring that students and faculty are engaged in developing and expanding community-university partnerships. Our project is, therefore, embedded in the intersection of DePaul University's Vincentian mission, and the significant resources it has put in place to engage faculty and students through curriculum with the community, with action research as an approach to producing scholarship that has an impact on social and ecological challenges. (https://mission.depaul.edu/vincentianidentity/documents/catholicidentity.pdf).

As illustrated by scholars from Dewey to Benson, universities are uniquely positioned to act as locally engaged education institutions (LEEI) where research and educational missions align with community goals to create partnerships channeled toward problem-solving in urban areas (See, 2010). Service learning is one way of local engagement. Defined as a pedagogical strategy in which students engage in community service that will enhance their understanding of course concepts and enable them to make contributions to their communities (Rhodes & Davis, 2001), the core aspirations of service learning are to increase civic engagement in students and narrow the distance between universities and communities (Benson et al., 2007; Morton & Saltmarsh, 1997). Well-designed service learning can have an array of student benefits, including greater content learning, critical thinking, civic responsibility, personal efficacy, democratic participation, tolerance, altruism, cultural awareness, leadership skills, and improved cross-cultural communication (Bringle & Steinberg 2010; Cashman & Seifer, 2008; Eyler & Giles, 1999).

Only half of the definition of service learning is tied to student learning. The other, equally important component, is contributing to communities. Scholars have recently expressed concern that many of the benefits of service learning accrue predominantly to students enrolled in service learning courses instead of the community (Mitchell, 2008). Care should be taken when designing service learning courses to avoid a service learning experience that suggests that communities have deficits that others can fill through service (Catlett & Proweller, 2011) or where students are veiwed as proto-experts, with the expertise to address community problems (Saltmarsh, Hartley, & Clayton, 2009), which does little to encourage long-lasting change in the community. To create a partnership where mutual goals are achieved requires communication and mutual respect of both partners (Lambert-Pennington et al., 2011). Indeed, Jacoby and Associates (2003) advised the careful identification of both unique and mutual goals of the university and their partners for each service learning project that they work on together coupled with regularly recording and assessing whether the goals are being achieved.

While not explicit when we created it, our project, a 3-week summer science immersion program, *Discover DePaul,* for low-income students, had an aspirational goal of transformative service learning in which participants would gain skills, knowledge, toolkits, and self-confidence necessary to continue working for change in their communities. As with many educational endeavors, *Discover DePaul* was meant to equip students with the ability to continue learning as they faced problems in their communities, and to use that knowledge to help work for change. Indeed, the learning in service learning results from the connections students make between their community experiences and course themes (Zivi, 1997). The mix of students in our program may have facilitated this connection. Our service learning course included 10 students from the community. These students had a rich knowledge of the community from their day-to-day experiences. Two undergraduate students from DePaul brought their college experience to the program, complemented with experiences from other service learning courses they had taken, or experience from other similar communities (as was the case for one of the undergraduates) in which they lived or worked. *Discover DePaul* was created with the idea of complementing this community experience with training in basic soil science, with particular emphasis on assessment of soil quality and health.

There is increasing scholarly interest in the design of college-level curricula and research projects that use community gardens, place-based education, and service learning to teach about basic soil science, soil quality, social justice, food justice and security, nutrition education, organic and urban agriculture, and green infrastructure, particularly in minority communities. For example, partly in response to declines in the number of students majoring in soil science and agronomy, several land-grant universities, including University of Tennessee–Knoxville, Ohio State University, North Carolina State University (Grossman et al., 2012) and University of Kentucky now offer place-based service learning soil-science courses with an emphasis on rural organic farming (Bhavsar, 2002). Non–land grant universities, including DePaul University and City College of New York, offer undergraduate soil-science and urban-farming-experiential-learning and service learning courses. Rosing and Odoms-Young (2015) highlight a number of community-university participatory research projects focused on improving access to and consumption of nutritious and healthy foods in low-income communities of color. All of these projects provide insight into the role of community-based participatory research (CBPR) in expanding food access, nutrition literacy, and health equity. One such project, conducted by Hatchett, Brown, Hopkins, Larsen, and Fournier (2015), studied the impact of urban agricultural programs on those who participated as gardeners and staff. They found that urban agriculture programs can improve food access in economically depressed African American communities, reconnect these communities with the land and their heritage, and foster entrepreneurship and leadership development. The Urban Nutrition Initiative, conducted by at the University of Pennsylvania, is a well-documented program whose overarching objective is to address nutrition-related issues that constitute an epidemic related to various health ailments in communities throughout the United States. Penn faculty and students in the UNI helped to design and deliver a nutrition and health curriculum in collaboration with teachers at a Philadelphia middle school, as well as conduct research on students' nutrient intake, diet quality, and the prevalence of overweight and obesity (Johnston et al., 2004). The undergraduate students become active learners who generate knowledge through community engagement. Aftandilian and Dart (2013) present several approaches for using garden-based service learning to address food justice and security; better educate undergraduate students about social justice, civic engagement, and personal responsibility for positive social changes; as well as strengthen campus and community relations.

## IMPORTANCE OF SOIL SCIENCE IN URBAN REDEVELOPMENT PROJECTS

Urban areas normally have complex land use histories, including the concentration of toxic substances in soil and water. Common examples of urban pollutants include sediment from development and new construction, oil, grease, toxic chemicals from automobiles, viruses and bacteria from failing septic systems, road salts, and heavy metals from industrial sources (Clark, Brabander, & Erdil, 2006; Ryan et al., 2004; ATSDR, 2000). Soils are the foundation for many ecological processes, including nutrient cycling habitat for biota, removal of toxic chemicals, and water filtration (Brady & Weil, 2010). While soils in urban areas are often disturbed by human activity, they play a vital role in the brown infrastructure of urban ecological systems in much the same way as vegetation serves as the green infrastructure (Pouyat, Yesilonis, Russell-Anelli, & Neerchal, 2007). Urban soils serve as sinks for heavy metals (Pouyat, Szlavecz, Yesilonis, Groffman, & Schwarz, 2010; Yesilonis, Pouyat, & Neerchal, 2008). Industrial activity, brownfields, landfills, and illegal dumping of contaminants often occur disproportionately in low-income communities, and contaminated soils then form the foundation for homes, playgrounds, parks, and gardens (Bullard, Mohai, Saha, & Wright, 2007; Hofrichter, 2002). EPA Toxic Release Inventory data indicate these pollution disparities are especially pronounced in metropolitan areas of the midwest (Mohai, Lantz, Morenoff, House, & Mero,, 2009).

Urban soils also contain high levels of nutrients (e.g., nitrates and phosphates) (Witzling, Wander, & Phillips, 2010). Nutrient-rich sediment runoff following rainfall can lead to eutrophication (Karavoltsos, Sakellari, Antonopoulou, Dassenakis, & Scoullos, 2010; Taebi & Droste, 2004) and seasonal hypoxia (Hey, Kostel, Hurter, & Kadlec, 2005) in receiving waters.

Sediment pollutant load is of particular concern in urban areas where drinking water sources are often receiving waters for stormwater runoff, and where treating this runoff at treatment plants is often costly. This is the case for the City of Chicago, where combined sewer systems (CSS) collect stormwater runoff, domestic sewage, and industrial wastewater in the same pipe. During periods of heavy rainfall or snowmelt, runoff volume in a CSS may exceed the CSS capacity. When this happens, gate and lock reversals can occur where a gate adjacent to the lock or the entire lock is opened to allow water to flow into Lake Michigan and relieve pressure on the CSS. Total overflow discharge in 2013 resulted in 10,710 million gallons of untreated sewage and stormwater flowing into Lake Michigan. Reducing this runoff prior to its arrival in the CSS is critically important. This can be done via infrastructure to safeguard the water supply (cf. the Deep Tunnel and Reservoir Project [TARP]); however, it may also require a network of small-scale projects to retain water at the parcel scale. To inform these parcel-scale solutions, an understanding of the underlying soil is foundational.

## SERVICE LEARNING, COLLEGE READINESS, AND COMMUNITY-BASED SOCIAL SCIENCE RESEARCH

A component of our project in Greater Grand Crossing centered on a better understanding of urban soils in low-income communities as a precursor to redevelopment (e.g., parks, vegetation, urban farms). A second and equally important component of the project was education. The larger question informing the program discussed here as well as our future research-education vision is How can we use science and education to help revitalize low-income communities, involve and mentor underrepresented students in applied scientific research in their communities, equip DePaul undergraduates with the skills to work with and learn from community partners, and compile and disseminate research results to help inform future environmental policy in the city of Chicago? The way this question is framed stresses partnership and collaboration. Without a true partnership, there is little hope that the impact of service learning will last any longer than the course. True partnerships also acknowledge the assets or skills partners bring to the table. Authentic partnerships require multidirectional learning, information sharing, and trust (Russell & Harshbarger, 2003). Such collaboration indicates working toward a mutual goal. It requires initial conversations to address how the skills and assets the different partners bring to the table can help change existing conditions in a way that addresses each organization's mission/goals either separately or as a cohesive whole.

This case study on *Discover DePaul* represents the initial step in responding to the above question and was implemented as a 3-week summer science immersion for high school students from the Gary Comer Youth Center (GCYC). *Discover DePaul*'s goal was assessing urban soil quality in underserved communities, using soil-quality assessment (Karlen, Andrews, Weinhold, & Doran, 2001; Karlen, Andrews, Weinhold, & Doran, 2003) as a portal into scientific exploration (both field and laboratory based) by youth in these communities, and ultimately, exposing these youth to opportunities in applied science and resources for college success. Cognizant of the aformentioned action research approches, we argue that working side-by-side with local partners is an effective way to help youth from the communities acquire the confidence necessary to serve as agents of change. Not only do these youth provide an example of what can be accomplished with hard work to inspire other youth, but they also know the community's assets and needs. With their community connections, these youth may be the best visionaries for what their neighborhood could become. Our approach seeks to develop a more transformational collaborative service learning by inviting these youth to participate in our endeavors (Mitchell, 2008).

To work toward this vision, *Discover DePaul* integrated service learning pedagogy by including a full-time undergraduate who served as a mentor, instructor, and lab assistant and another undergraduate who served as a part-time mentor and lab assistant for the program. The immersion nature of the program allowed for more depth in the interactions between the high school students and the undergraduates than would have been possible in a semester-long service learning course. The time devoted to interaction between the high school students and undergraduates is crucial to both mentorship and the service learning component. In an assessment of student complaints on service learning at DePaul, one theme for improvement was a desire to spend more time with community residents to better understand the people and the issues in the community (Rosing, Reed, Ferrari, & Bothne, 2010). Grossman et al. (2012) also found that students faced signficant challenges interacting with community members during a pilot service learning program (the Community Food Security Scholars program) at North Carolina State University.

For the high school students, this course included components of the honors seminar approach which works by increasing the expectations that students have for their own success and which has increased minority retention in mathematics and science (Pinto, 1997 as cited in Bain, 2004). These honors-style activities are based on the finding that student awareness of negative stereotypes can lead to performance anxiety that can depress academic performance (Aronson et al., 1999; Croizet, Désert, Dutrévis, & Leyens, 2001; Spencer, Steele, & Quinn, 1999; Steele & Aronson, 1995), termed *stereotype vulnerability*. A woman who does well in mathematics in high school and the early college years or a black student who excels in any subject may dream of a future in the field, but it is those dreams that stimulate what Steele (1999) called vigilant worry that their future will be compromised by society's perception and treatment of their group. Indeed, stronger students are more susceptible to stereotype threat (Bain, 2004; Shih, Pittinsky, & Ambady, 1999; Steele, 1999). Giving students extra help in a remedial class therefore reinforces the student's perception that society believes that they will not succeed and negatively impacts performance, even if the student is more than competent. Conversely, an honors seminar where students encounter the expectation that they will succeed has been shown to be strongly significant in promoting minority student retention and success in mathematics and science programs (summarized in Bain, 2004). Therefore, the service learning pedagogy and the honors seminar approach were combined to provide an intensive service learning experience for the DePaul undergraduates while promoting the college readiness and expectations for career success in the high school students.

## THE DEPAUL UNIVERSITY AND GARY COMER YOUTH CENTER PARTNERSHIP: A HISTORY OF INNOVATION

This program was implemented collaboratively with our partner institution the Gary Comer Youth Center (GCYC). The GCYC is located in the predominantly African American neighborhood of Greater Grand Crossing on Chicago's south side, where approximately 78% of residents live below the poverty level. Members have the opportunity to contribute to their community through programs that focus on civic responsibility as well as personal and community development. One example is the Green Teens program where high school students use the Center's green roof and new urban garden as outdoor laboratories to enrich their educational experience and prepare them for college and careers. There is a strong educational mission at GCYC with a clear goal of 100% of high schoool students graduating from college and career ready. For the history of the adjacent charter high school (Gary Comer College Prep, which opened in August 2008), the first two cohorts of seniors all applied and were accepted to colleges and universities. The mission of GCYC is strongly tied to DePaul University's mission, which includes a "commitment to the education of first generation college students, especially those from the diverse cultural and ethnic groups in the metropolitan area" (DePaul University, n.d.a, para. 1). Interpreted broadly, DePaul's mission includes creating opportunities to increase the success of first-generation students, whether those programs occur before or during college.

DePaul has a long and successful history of implementing service learning pedagogy through support of the Steans Center for Community-based Service Learning. The Center acts as a liason between the university and communities through Chicago through its support of student learning, community development and engaged teaching and scholarship. Mutually beneficial relationships with community organizations are important, as the Center's mission states, "to engage DePaul students in educational opportunities grounded in Vincentian values of respect for human dignity and the quest for social justice" (DePaul University, n.d.b). Indeed, GCYC, our community partner in this project, has had a long-standing relationship with the Steans Center director. According to Marji Hess, Garden Manager at the Gary Comer Youth Center,

> The partnership between DePaul and GCYC actually goes back to 2007. Just shortly after we opened, Howard Rosing [director of DePaul's Steans Center] approached us and wanted to have an anthropology class come down and take a look at the food system in the community. And the idea was if we taught young people how to grow and cook food, how would that impact their lives? That was a fascinating conversation and we really appreciated that a university of DePaul's caliber—we were a youth center not even open 6 months—they could already see the potential in what we hoped we'd be able to accomplish. So from that time, we have moved to this active partnership, we have two interns every summer through the Environmental Science Department and the Steans Center, and this past summer we had this opportunity to really dig deeper and do *Discover DePaul*. That was a 3-week intensive program where 10 of our young people went to DePaul for 3 weeks and learned how to take soil samples, run the tests and then present them. But they also got to learn what it's like to go to college. And that so ties to our mission of our young people ready to graduate from high school ready for college and careers.

## DISCOVER DEPAUL: AN OVERVIEW

In 2013 we received funding from the Vincentian Endowment Fund at DePaul University to implement a project whose overarching goal was to strengthen the partnership between DePaul and the Gary Comer Youth Center. In collaboration with educators at GCYC, we developed a 3-week program called *Discover DePaul*.

### Week One

Green Teens spent the first week developing the scientific skills necessary to assess soil resources using GCYC's urban garden as a training site. This week included an introduction to the scientific method. Green Teens were introduced to experimental design and hypothesis testing using various interactive hands-on activities. During this week they worked under the tutelage of DePaul environmental studies major Xochyl Perez, as well as Dr.

Klimas and Dr. Montgomery, to collect soil samples from GCYC's rooftop garden and adjacent urban farm. Green Teens analyzed these soil samples for nutrients using equipment in the environmental soil science laboratory located in the Department of Environmental Science and Studies at DePaul during week two of *Discover DePaul.* In the first week they became acquainted with a suite of site-based tests routinely conducted on soils, including counting the number of earthworms, quantifying soil texture, soil color, soil structure, and stability. Once they acquired these basic skills, they conducted similar soil quality assessments on soils found in nearby vacant lots owned by GCYC. In addition, they took two field trips to nearby urban green spaces (including the Midewin National Tallgrass Prairie near Joliet, Illinois; North Park Nature Center in the city of Chicago; Heller Nature Center; and Prairie Wolf Slough wetland, both in Highland Park, Ilinois, to look at differences in local soils.

### Week Two

During the second week of *Discover DePaul,* the Green Teens were based at DePaul's Lincoln Park campus, where they conducted chemical analyses of the soil samples collected from the GCYC urban farm, from other GCYC properties, and from natural areas visited in the first week. This testing was done under the guidance of Xocyl Perez, Dr. Klimas, and Dr. Montgomery in the environmental soil science lab. Green Teens arrived at DePaul via public transportation and spent the day alternating between soil analysis in the lab work and various trips and lectures on and around campus. Trips included a walk around the Lincoln Park Zoo with wildlife biologist Mason Fidino, a tour around the Peggy Notebart Nature Museum, a campus tour including a tour of DePaul's rooftop gardens and urban farm, and a visit with DePaul Art Museum's curator Laura Fatemi. Soil laboratory tests from week two included total lead, carbon/nitrogen ratio of soil, nitrate, and phosphate.

### Week Three

During the third week of *Discover DePaul,* the Green Teens finished conducting soil analyses, analyzed the data using graphical and basic statistical techniques, and prepared an oral presentation on their research. This week also included a series of 2-hour lunchtime discussions with topics including financial aid, library research, and student life. Speakers included the Dean of Students and representatives from several organizations and offices, including DePaul Men of Color, Academic Advising, the Career Center, the

Office of Multicultural Student Success, and University Ministry. Professors were also invited to present lunch time discussions directed at the the intended majors of the Green Teens. Vijay Pendakur and Fr. Memo Campuzano each spoke to students about their personal experiences in overcoming obstacles as well as the support available at DePaul University. Additional activities included a student panel discussion on research in the College of Science and Health, tours of science faculty research labs, and a tour of DePaul's downtown campus (including the gaming labs for the students interested in DePaul's game design major).

The 3-week itinerary was developed to balance science learning, the use of scientific knowledge to answer local questions of interest, and an introductory college experience for the Green Teens. This is an extension of an annual one-day summer visit to DePaul for all 40+ Green Teens organized by the Steans Center. We consider the introductory college experience an important part of the program since it gives teens the opportunity to envision themselves at DePaul. In addition, it builds the capacity of DePaul's science, technology, engineering, mathematics (STEM) pipeline, which was established by DePaul's Office of STEM Studies. Green Teens involved in the one-day trip have found the experience transformative. Indeed, one former student said, "It was the best day of my entire summer."

## KEY ASPECTS OF THE PARTNERSHIP

Project implementation was successful in large part due to the support of many individuals at GCYC. The Green Teens who formed the *Discover DePaul* cohort were paid by GCYC with grant funds. GCYC provided meals for the first week of *Discover DePaul* (when we were located at Comer or Chicagoland natural areas). They also provided transportation for all field trips as well as transit cards for travel to and from DePaul in the later 2 weeks of *Discover DePaul*. Additionally, we had the assistance of Comer staff member Cora Marquez, who was also a former Steans Center intern. This was a tangible demonstration of the value of the program on the part of both DePaul and GCYC. It also gave the students security when traveling to and from DePaul, and allowed the professors to focus on the academics of the course. Cora's enthusiasm and rapport with the students was invaluable. In addition, open communication during the planning and implementation process was critical, allowing us to customize some of the speakers to match student interests.

Students applied to, and were selected for, this program by GCYC. In order to apply, students were asked to submit a brief statement of interest in the program, one recommendation letter, and evidence that they had completed 2 years of science with a grade of B or higher. The program included high expecations coupled with mentoring by DePaul students, lead faculty, GCYC staff, as well as multiple opportunities for students to interact with others at DePaul.

## SURVEY PROCEDURE, RESULTS, AND DISCUSSION

Our intensive education module focused on a scientific exploration of soil was developed to help a small group of youth gain skills necessary for careers and future studies, perhaps in the natural and health sciences, as well as giving them a vision of how to use their skills to improve their community. One of the key components of this program was the direct invovement of those most affected by the challenges of contaminated soil, thus training them in the science necessary to understand the problem as well as the research skills to work toward solutions. Following completion of *Discover DePaul,* students completed two surveys. The first had students respond to 14 questions using a Likert scale where 1 = strongly disagree, 2 = disagree, 3 = agree, 4 = strongly agree (Table 7.1). In addition, the first survey asked three open-ended questions: (a) I especially liked... (b) What suggestions would you make to improve Discover DePaul in the future? (c) Do you think that Discover DePaul is a worthwhile activity for future green teens? All students completed this for a sample size of $n = 9$. Significance ($p$) of the difference of the average student response for Likert-scored questions from a neutral (neither agree nor disagree) average of 2.5 was determined by a Wilcoxon signed ranks test.

The second survey had students check different boxes indicating their self-assessment of whether their participation in *Discover DePaul* led to a gain or improvement in listed employability skills (Table 7.2). This survey also included open-ended questions, including (a) Do you believe you have gained anything such as education, job skills, college readiness, etc. from being a part of this program?; (b) What would you have done differently or changed about this program/experience?; (c) Would you recommend this program to your friends? Why or why not?; (d) Please write any final reflections or comments you would like to make about your experience or the program; and (e) What if any of the following components of Discover DePaul do you think prepared you for success in college or a job? Seven of the nine students completed the second survey.

## STUDY LIMITATIONS

Before we elaborate on the results, it is important to both put our results in perspective and bound their generalizability. With only 10 students selected

**TABLE 7.1 Student Responses To Survey Questions**

| Survey Question | Average | *SD* | *P* | % of Students Who Agree | Avg. Student Response |
|---|---|---|---|---|---|
| After participating in *Discover DePaul*, I think I could design other research experiments. | 3.2 | 0.4 | 0.006 | 100 | agree |
| I learned to do the type of research that is required of research scientists. | 3.7 | 0.5 | 0.007 | 100 | agree |
| *Discover DePaul* helped me understand real-world issues related to soil and agriculture. | 3.4 | 0.5 | 0.007 | 100 | agree |
| I am more excited about science than before. | 3.1 | 0.6 | 0.025 | 89 | agree |
| I learned something from other people in my research team. | 3.4 | 0.5 | 0.007 | 100 | agree |
| The program was more difficult than most of my science labs. | 3.1 | 0.6 | 0.025 | 89 | agree |
| I feel good about my group's final presentation. | 3.1 | 0.3 | 0.005 | 100 | agree |
| I learned something from seeing other people's final projects. | 3.4 | 0.5 | 0.007 | 100 | agree |
| During *Discover DePaul*, I was bored more than in most classes. | 1.7 | 1.1 | 0.007 | 78 | disagree |
| I learned more about the college application process from participating in *Discover DePaul*. | 3.3 | 0.7 | 0.021 | 89 | agree |
| I learned more about available financial aid opportunities from participating in *Discover DePaul*. | 3.1 | 0.6 | 0.025 | 89 | agree |
| I learned more about resources available for student success from participating in *Discover DePaul*. | 3.3 | 0.5 | 0.007 | 100 | agree |
| I enjoyed *Discover DePaul*. | 3.8 | 0.4 | 0.006 | 100 | agree |
| I feel more ready for work in a college environment after participating in *Discover DePaul*. | 3.4 | 0.5 | 0.007 | 100 | agree |

*Note:* 1 = Strongly Disagree, 2 = Disagree, 3 = Agree, 4 = Strongly Agree, for a Sample Size of $n = 9$. Significance ($p$) of the difference of the average student response from a neutral (neither agree nor disagree) average of 2.5 was determined by a Wilcoxon Signed Ranks Test.

**TABLE 7.2 Student Responses Self-Assessing Whether They Gained or Improved in Different Content Areas (*N* =7 for All Categories)**

| Employability Skills Gained/ Improved | % of Students Reporting Gain/Improvement |
|---|---|
| *Active listening* | *100* |
| Career readiness | 71 |
| *College readiness* | *100* |
| Communication | 71 |
| Cooperation | 71 |
| Creating consensus | 14 |
| Decision making | 71 |
| Follow through | 57 |
| Leadership | 71 |
| *Participation with peers* | *100* |
| Planning | 57 |
| Presentation | 86 |
| Prioritizing goals | 86 |
| Prioritizing tasks | 71 |
| Problem solving | 71 |
| Public speaking | 85 |
| Reflection | 71 |
| *Teamwork* | *100* |
| *Time management* | *100* |
| Writing | 43 |

from multiple Chicago-area schools, attributing current and future success of students to participation in *Discover DePaul* is impossible. Indeed, we cannnot disaggregate *Discover DePaul* from students' larger academic experience. Pre- and postsurveys would have also been useful in determining the success of this program, as would assessments of improvement in short-term and long-term retention of soil quality concepts and college services for success. The relationship between short-term and long-term retention is still somewhat unsettled. For example, in their comparison of short-term retention versus long-term retention both in lecture combined with discussion courses and cooperative learning courses, Morgan et al. (2000) reported minimal differences in long-term retention. However, Balleck (2006) and Pérez-Sabater, Montero-Fleta, Pérez-Sabater, and Rising (2011) reported increased long-term retention compared to control groups when employing active and collaborative learning strategies. *Discover DePaul* employed cooperative learning strategies which have been shown to motivate students' interest and improve their retention of key ideas (Abu & Flowers, 1997; Rosetti & Nembhard,1998). As our project was a pilot project not tested for general applicability in different situations or to be rolled out at a larger scale, our results are limited only to students' perceptions. Perceptions are powerful. Research on identity theory has shown that students who perceive themselves to have a positive self-image, even if innacurate, have the necessary confidence needed to engage in and succeed in new challenges (Baumeister 1998; Crosnoe, Riegele-Crumb, & Muller, 2007; Sedikides, 1993).

## SUCCESSES

Postsurvey results from this program highlight multiple metrics of success based on student self-evaluations. *Discover DePaul,* co-led by ecologist Christie Klimas and DePaul soil scientist Dr. James Montgomery, was extremely well-received by students and led to students self-reporting higher college readiness. Before we describe the student responses, it is critical to mention that this was a transformational learning opportunity for all involved in the project, including both lead faculty. As eloquently stated by Dr. James Montgomery,

> This summer *Discover DePaul* project was a very eye opening experience for me. I think as I mentioned in this article about taking them out to one of my longtime research sites and asking them to listen and smell and when we reconvened and I asked how is this environment different from where you grew up, and this young woman said I didn't hear any gunshots. That just floored me like a ton of bricks.

Both Klimas and Montgomery are more typical of the student demographic that is normally enrolled in service learning courses. These students often come from priviledged backgrounds with a majority being White and middle-class students that work in lower income communities of color (Green, 2001). Indeed, this was the case for both of the program's instructors. Dr. James Montgomery detailed the collaborative nature of our approach, which we termed our "train-the-trainers" model:

> This drives home the aspect of working with an anchor institution in a low-income minority community that has a lot of needs. And what can we do to help meet those needs in a way that's not intrusive, that's not about a bunch of white people coming in and telling them how to do things, but where they have a seat at the table and help to figure out what the alternate solution was.

As is evident in Table 7.2, the majority of the students who completed the second survey reported a gain or improvement in 18 of the 20 employability skills. Exceptions were "creating consensus" and "writing," neither of which were part of the core competencies addressed in *Discover DePaul*. Our GCYC partners also viewed the project as a success. Marji Hess stated,

> The idea is the mission of the Gary Comer Youth Center is for our students to graduate from high school ready for college and careers. The *Discover DePaul* program made that real; it was no longer just an idea. This was going to college, being on a college campus, and seeing what college kids looked like. And thanks so much to what Christie and Jim were able to do; it was a broader experience because it was about financial aid and meeting the academic services department and getting to look all around the different science departments. So for us that was the biggest part, but the other side is that the garden department is very interested in launching green careers, not just going to college. If we had our way, we'd want to inform that green stripe of college careers. The greatest takeaway is the confidence our young people gained about pursuing a science-based career. The thought of "I can do this, I can see myself doing this, I have done it, I learned this skill, I presented this skill, I could be among peers, I was respected, and I could contribute to the conversation." So the confidence that they gained was huge.

Students reported similar gains from participating. In response to the question, "Do you believe you have gained anything such as education, job skills, college readiness, etc. from being a part of this program?" students listed multiple gains (a complete list of student responses is in the Appendix ). These comments include references to an increase in content knowledge of soil science: "I've been educated about environmental science (more specifically soil science). I've encountered different people that made different career choices and got their advice and I've learned MANY things about college life and preparation for it. In other words, yes" and "Yes, I believe I gained college readiness from having a taste of lab research and how it felt to be a science major at DePaul." Both of these comments also include references to increasing student knowledge about college life/college readiness. Another student comment also mentioned college readiness coupled with a comment on a better understanding of the structures in place at colleges to help students succeed: "I've learned job skills and college readiness and how it was on campus and the support system you'll have." We even had one student comment on the difference between a degree and a career, indicating that this program encouraged or strengthened the time students spent thinking about postcollege life (or college as a long-term step). "I gained a lot more interest in environmental science and I'm more interested in getting a degree as well as a career."

The gains in education and college readiness were mirrored in their responses to the Likert-scale questions (Table 7.1). All student responses were significantly different from a neutral score of 2.5 as determined by a Wilcoxon signed ranks test. Students agreed that their science and soil content knowledge was sufficient to design and conduct research experiments (Table 7.1, Q1–3), though future programs could assess this by having students develop follow-up experiments. Additionally, their enthusiasm for science increased after *Discover DePaul* (Table 7.1, Q4). Their responses also indicated an increase in college readiness, including applying for financial aid and support structures available at college (Table 7.1, Q11–12, 14).

Students were asked open-ended questions to reflect on how *Discover DePaul* helped prepare them for college and job success. What if any of the following components of Discover DePaul do you think prepared you for success in college or a job? Five students highlighted the scientific process: (a) "experiments," (b) "Doing the in-depth experiments prepared me for college. I got to see how differently they are conducted and more rigorous than high school," (c) "Researching and conducting different experiments with a big group helped the most," (d) "Going through the full process of what they would do," and (e) "The presentation, research, and lab work prepared me mostly for majoring in science at DePaul/college." Two students highlighted the group work, mentioning talking in a group and the ability to work with different people. Two other students highlighted the importance of the support structures, including: "learning about financial aid," and "I would have a support system and how it helps you prepare for real life." These comments holistically include what we considered the critical components of *Discover DePaul*. Understanding the scientific process and how to use science in applied situations was one goal. Most high school curricula do not effecively employ scientific inquiry-based teaching (Crawford, 2007). Additionally, soil science content is rare (Collins, 2008). Using applied science to inform redevelopment of vacant properties was the model around which we developed the course.

The service learning component was also key. Delbanco (2012, p. 177) concludes *College: What It Was, Is, and Should Be* with the statement that "College should [be]...where they [students] discover that self-interest need not be at odds with concern for one another. We owe it to posterity to preserve and protect this institution. Democracy depends on it." Indeed, public investment in science is predicated upon a collective expectation of a return of knowledge and technology for the benefit of society (Lubchenco, 1998). However, this return is not always immediately apparent, nor does it involve or reach those to whom the results would be most valuable. DePaul's urban setting provides a rich opportunity to connect knowledge with those communities for whom it would be most valuable. Service learning coupled with community participation and engagement has the potential to bridge the divide between science and its applications (Bringle, Games, & Malloy, 1999) while having the additional benefit of forming new scientists. One of the students in the program highlighted the service connection in a postprogram interview approximately 6 months after the conclusion of *Discover DePaul*:

> In my way, I want to give back to Chicago since I spent the majority of my life here. Not here per se but some of the more impoverished neighborhoods, like Englewood, so 10, 15, even 20 years down the line, I could say I grew up around here and wow it's looking nicer. Don't look at it like dirty and squalor; we have to try to make it look as beautiful as somewhere else; we have to take the initiative to get involved and give back, because in the long run, it will benefit everyone. Chicago's going to be nature friendly.

This comment typifies the desired response from the change model of service learning, which "facilitates students' critical reflection on social problems and considers the role they can play in engaging with communities to transform and shape a different social landscape" (Catlett & Proweller, 2011). In other words, the comment indicates that *Discover DePaul* has promise as a program where those who participate gain the skills, knowledge, toolkits, and self-confidence necessary to continue working for change in their communities. Future similar programs could explicitly test whether this was true for the majority of participants. While not all students mentioned this, the hope is that *Discover DePaul* set the stage for the teens to visualize what could be in their neighborhoods. This potentially transformational aspect was also mentioned by Marji Hess:

> This has been a very asset-based experience. A lot of people only come to this community and see the deficits. But DePaul as an organization came to us and saw the assets in the community and helped our young people to see that. And the other part is persistence. Persistence working with us models to our young people what persistence is. If you are an African American male who's a ninth grader, there's a 4% change you will end up graduating from college. If you're a female, there's a 7% chance.... They saw themselves at such pivotal times in their lives when they're defining who they are; DePaul saw assets in them, you gave them that idea that we persist. You stay with these projects. You come back. It's not just "Okay we're done," but here's what's next, we're in it with you, how happy you were for them when they were talking about their college choices.

And reciprocally, it was easy to view the *Discover DePaul* students as current and future agents of change, a large part of the model of action research (AR). Greenwood and Levin (1998) state that

> AR is a form of research that generates knowledge claims for the express purpose of taking action to promote social change and social analysis.... AR aims

to increase the ability of the involved community or organization members to control their own destinies and to keep improving their capacity to do so. (p. 6)

## AREAS FOR IMPROVEMENT

Students were overwhelmingly positive about *Discover DePaul* in their comments and reviews (Tables 7.1, 7.2, & Appendix) and their comments supported the program, recommending few changes. However, there are changes that we think are important for future programs which complement student suggestions. One student mentioned that "Getting to DePaul was very tiring, however, I enjoyed the program so much that I didn't mind." Group transportation was only provided for the field trips. During the 2 weeks at DePaul, the students took public transportation. Due to repair work on the trains, this took over one hour each way, and was exhausting. Indeed, it was one of the priority items that came out of discussions on how to improve the experience for all. Complaints about transportation being time consuming and detracting from an otherwise meaningful experience have been found in other service learning research (Conner, 2004; Rosing et al., 2010). Other students mentioned a desire for more lab time: "I would've changed the amount of lab time. I would like more lab time to use the equipment during experiments." This could be done in the context of the 3-week program, but there is another way to continue the learning for those students who were ready to engage more fully in the research. One way is to add an optional research component that students could elect to participate in following *Discover DePaul.* The other idea we discussed was to make this program annual or every other year. This would allow students the opportunty to participate again, but in the role of mentor. They would also then be in charge of more of the instruction, giving them more lab time and a more in-depth understanding of the soil-quality tests.

While we did not complete a pre- and postprogram assessement for *Discover DePaul,* the postprogram evaluations are extremely encouraging and recommend subsequent iterations of the program. With future iterations, we would conduct pre- and postassessments for both content knowledge and college readiness (possible appropriate methodologies are summarized in Newman, Dantzler, & Coleman, 2015). Programs of this type, while only one factor affecting student success in college, should be tested to determine whether they play a significant role in contributing to student success. This is particularly important for minority groups, like the students who participated in *Discover DePaul.* According to Steele (1999), throughout the 1990s the national college-dropout rate for African Americans has been 20% to 25% higher than that for whites. Among those who finish college, the grade-point average of black students is two thirds of a grade below that of whites. In addition, testing to see whether the honors seminar approach is effective prior to entering college is an important part of the mission of both institutions that participated in this research.

## CONCLUSIONS AND FUTURE DIRECTIONS

Programs like *Discover DePaul* may have potential for encouraging collaboration among students, teachers, and community partners, creating civic and social responsibility for both university and high school students, stimulating resiliency for youth in economically distressed communities, and for stimulating interest in the STEM disciplines and careers. One of the key components of this program was the direct invovement of those most affected by these challenges, thus training them in the science necessary to understand the problem as well as the research skills to work toward solutions. In addition to the science component of the program, *Discover DePaul* addressed themes relevant for the teaching and learning of K–16 science and service learning combined with an honors seminar approach with the intention of increasing the expectations that students have for their own success. Preparing students for college and career success addressed a key concern of GCYC, as well as the concerns and needs of the community.

*Discover DePaul* provided the foundation for a subsequent community-improvement project that also coupled education and soil quality analysis to revitalize vacant lots near the GCYC. In this project, a group of DePaul students are using soil-quality analysis to determine appropriate types of green infrastructure (GI) redevelopment for GCYC-owned vacant lots. This information was used in a DePaul landscape design course to create models of GI development options. We are in the process of applying for funding to use this information in a participatory community-decision-making process to design a community space and implement the community's design. These and other prospective projects are not simply the product of studying soil in Chicago neighborhoods. We recognize both practically and figuratively that soils are in the city and that the "the city" is made up of people who have aspirations to improve their lives, especially in areas experiencing historical social and political marginalization, economic distress, and environmental problems. Youth in those neighborhoods are their future, and we intend to further explore how a science-based education informed by a pedagogy involving action-based research can ensure their success.

## APPENDIX
## Student Responses to Open-Ended Questions

### *Do you believe you have gained anything such as education, job skills, college readiness, etc. from being a part of this program?*

"I think I have gained education skills from doing college-level experiments."

"I've been educated about environmental science (more specifically soil science). I've encountered different people that made different career choices and got their advice and I've learned MANY things about college life and preparation for it. In other words, yes."

"I've learned job skills and college readiness and how it was on campus and the support system you'll have."

"Yes, I believe I gained college readiness from having a taste of lab research and how it felt to be a science major at DePaul."

"I gained a lot more interest in environmental science and I'm more interested in getting a degree as well as a career."

"Yes, I believe that I gained college readiness skills because now I kind of know what to expect when I get there. I know how a campus looks."

"Yes, I learned that you have to start early."

### *What would you have done differently or changed about this program/experience?*

"I would've changed the amount of lab time. I would like more lab time to use the equipment during experiments."

"Nothing... They fed us, taught us various things, [?] us and fed us. Did I mention that they fed us?"

"Absolutely nothing. Everything was wonderful and how it changed my life in some ways I took away a lot from it."

"I wouldn't probably change anything because it was an amazing and rare experience that I won't ever forget."

"Getting to DePaul was very tiring, however I enjoyed the program so much that I didn't mind."

"One thing that I would have changed is that I think that the tours should have been a little more interesting because at some times they got a little boring. But other than that everything else was good to me."

"The amount of time we spent in the lab."

### *Would you recommend this program to your friends? Why or why not?*

"I would highly recommend this to my friends because I thoroughly enjoyed the program and believe my friends would too."

"YES! To anyone who is interested in environmental science or wants to prepare for college, or even those who want to explore different career options, this program is for you."

"Yes, to anyone who is interested in the studies and sciences and who is really dedicated to it not just for many."

"I would only recommend this program to my friends who are interested or have respect for science. Also, I'd recommend this to friends who I know that are hard working."

"I would definitely recommend this program to my friends but only the ones that enjoy science because it's great exposure."

"Yes, because they would have a chance to be able to get the chance to experience how college is going to be because everyone should want to go. They get a chance to work on different projects and have fun while they are doing it."

"I would because I learned so much in the 3 weeks I've been in the program."

***Please write any final reflections or comments you would like to make about your experience or the program.***

"N/A—It was fine!!!"

"It was one of the most immersive experiences of my life. And they fed us."

"All I have to say is it was a good experience to prosper more."

"The professors were absolutely amazing, caring, and friendly this summer! Shout out to Dr. Monte and Dr. Klimas!"

"My experience in this program will not be forgotten. I found it all very interesting. This increased my interest to study environmental science."

"In enjoyed my time at DePaul. They helped me out a lot with getting ready for college. The people that I got a chance to meet were very nice."

"I wish I can get another lab coat."

What, if any, of the following components of Discover DePaul do you think prepared you for success in college or a job?

"Researching and conducting different experiments with a big group helped the most."

"Talking in a group."

"The presentation, research, and lab work prepared me mostly for majoring in science at DePaul/college."

"learning about financial aid."

"experiments"

"Doing the in-depth experiments prepared me for college. I got to see how differently they are conducted and more rigorous than high school."

"Going through the full process of what they would do."

"I would have a support system and how it helps you prepare for real life."

"How to work with different people."

***I especially liked . . .***

"the labs and the lab coats."

"our powerpoint presentation."

"when we got to tour DePaul and ate Anette's ice cream."

"learning about soil."

"the trips."

"the college of computing and digital media (the awesome gaming labs J)"

"working in the labs like actual scientists."

"the professor and the experience."

***What suggestions would you make to improve Discover DePaul in the future?***

"Have more food choices. Have more lab time."

"Nothing."

"None"

"None."

"That it is not so long."

"More experiments!"

"I would like better transportation."

"Nothing just for it to be longer."

Do you think that continuing Discover DePaul is a worthwhile activity for future Green Teens?

"YES!!!!"

"yes"

"Yes"

"YES."

"yes"

"Yes, highly recommend *Discover DePaul* stays for future green teens to enjoy."

"Yes. I strongly recommend it."

"Most deff; pursue it if you love science and just [?]."

## REFERENCES

Abu, R. B., & Flowers, J. (1997). The effects of cooperative learning methods on achievement, retention and attitudes of home economics students in North Carolina. *Journal of Vocational and Technical Education, 13*(2). Retrieved from http://scholar.lib.vt.edu/ejournals/JVTE/v13n2/Abu.html

Aftandilian, D., & Dart, L. (2013). Using garden-based service learning to work toward food justice, better educate students, and strengthen campus-community ties. *Journal of Community Engagement and Scholarship, 6,* 55–69.

Agency for Toxic Substances and Disease Registry (ATSDR). (2000). *Lead toxicity* [ATSDR Publication No. ATSDR-HE-CS-2001-0001]. Washington, DC: U.S. Department of Health and Human Services.

Aronson, J., Lustina, M. J., Good, C., Keough, K., Steele, C. M., & Brown, J. (1999). When White men can't do math: Necessary and sufficient factors in stereotype threat. *Journal of Experimental Social Psychology, 35,* 29–46.

Asselin, M., & Brisbois, M. (2015). Enhancing global service learning experiences of baccalaureate students through critical reflective inquiry: A qualitative study. *Nursing Research, 64*(2), E45.

Balleck, B. (2006). *Teaching for the future: Experiential learning and the use of new strategies and techniques available to the international studies classroom.* Paper presentened at the International Studies Association Annual Meeting.

Bain, K. (2004). *What the best college teachers do.* Cambridge, MA: Harvard University Press.

Baumeister, R. (1998). The self. In J. Howard & P. Callero, P. (Eds.), *Handbook of social psychology* (pp. 780–839). Cambridge, England: Cambridge University Press.

Benson, L., Harkavy, I., & Puckett, J. (2007). *Dewey's dream: Universities and democracies in an age of education reform.* Philadelphia, PA: Temple University Press.

Bhavsar, V. M. (2002). Certified organic farming principles and practices: A course linking farmers and university students. *Journal of Natural Resources and Life Sciences Education, 31,* 20–24.

Brady, N. C., & Weil, R. R. (2010). *Elements of the nature and properties of soils* (3rd ed.). New York, NY: Prentice-Hall.

Bringle, R. G., Games, R., & Malloy, E. A. (1999). *Colleges and universities as citizens.* Needham Heights, MA: Allyn & Bacon.

Bringle, R. G., & Steinberg, K. (2010). Educating for informed community involvement. *American Journal of Community Psychology, 46,* 428–441.

Bronfenbrenner, U. (1979). *The ecology of human development: Experiments by nature and design.* Cambridge, MA: Harvard University Press.

Buff, S. M., Jenkins, K., Kern, D., Worrall, C., Howell, D., Martin, K., . . . Blue, A. (2015) Interprofessional service learning in a community setting: Findings from a pilot study. *Journal of Interprofessional Care, 29,* 159–161.

Bullard, R. D., Mohai, P., Saha, R., & Wright, B. (2007, March). *Toxic wastes and race at twenty, 1987–2007.* Retrieved from http://www.ejnet.org/ej/twart.pdf

Butin, D. (2015). Dreaming of justice: Critical service learning and the need to wake up. *Theory into Practice, 54,* 5–10.

Cashman, S. B., & Seifer, S. D. (2008). Service learning: An integral part of undergraduate public health. *American Journal of Preventive Medicine, 35,* 273–278.

Catlett, B. S., & Proweller, A. (2011). College students' negotiation of privilege in a community-based violence prevention project. *Michigan Journal of Community Service Learning, 18,* 34–48.

Clark, H. F., Brabander, D. J., & Erdil, R. M. (2006). Sources, sinks, and exposure pathways of lead in urban garden soil. *Journal of Environmental Quality, 35,* 2066–2074.

Collins, M. E. (2008). Where have all the soils students gone? *Journal of Natural Resources and Life Sciences Education, 37,* 117–124.

Conner, D. B. (2004). The effects of course-related service projects in a child development course. *College Student Journal, 38,* 62–471.

Crawford, B. A. (2007). *Learning to teach science in authentic settings: Moving students to the inner circle.* New York, NY: American Educational Research Association Conference.

Croizet, J.-C., Désert, M., Dutrévis, M., & Leyens, J.-P. (2001). Stereotype threat, social class, gender, and academic under-achievement: When our reputation catches up to us and takes over. *Social Psychology of Education, 4,* 295–310.

Crosnoe, R., Riegele-Crumb, C., & Muller, C. (2007). Gender, self-perception, and academic problems in high school. *Social Problems, 54*(1), 118–138.

Delbanco, A. (2012). *College: What it was, is, and should be.* Princeton, NJ: Princeton University Press.

DePaul University (n.d.a). *DePaul University mission statement.* Retrieved from https://offices.depaul.edu/mission-and-values/about/Documents/DePaul%20University%20Mission%20Statement.pdf

DePaul University (n.d.b). *About.* Retrieved from http://steans.depaul.edu/About/IrwinWSteans/Mission

Dewey, J. (1888). *The ethics of democracy. Pragmatism: A reader.* New York, NY: Louis Menand Vintage.

Emery, F. E., & Trist, E. L. (1973). *Towards a aocial ecology: Contextual appreciation of the future in the present.* London, England: Plenum Press.

Eyler, J., & Giles, D. E., Jr. (1999). *Where's the learning in service learning.* San Francisco, CA: Jossey-Bass.

Freire, P. (1970). *Pedagogy of the oppressed.* New York, NY: Continuum.

Furco, A. (2001). Advancing service learning at a research university. In M. Canada & B. W. Speck (Eds.), *Developing and implementing service learning programs* (pp. 67–78). San Francisco, CA: Jossey-Bass.

Greenwood, D. J., & Levin, M. (1998). *Introduction to action research: Social research for social change.* Thousand Oaks, CA: Sage.

Grossman, J., Sherard, M., Prohn, S. M., Bradley, L., Goodell, L. S., & Andrew, K. (2012). An exploratory analysis of student-community interactions in urban agriculture. *Journal of Higher Education Outreach and Engagement, 16,* 179–196.

Hatchett, L., Brown, L., Hopkins, J., Larsen, K., & Fournier, E. (2015). Something good can grow here: Chicago urban agriculture food projects. *Journal of Prevention & Intervention in the Community, 43,* 135–147.

Hey, D., Kostel, J., Hurter, A., & Kadlec, R. (2005). Nutrient farming and traditional removal: An economic comparison. *Water Intelligence Online.* Retrieved from www.iwaponline.com/wio/2005/06/wio200506WF03WSM6CO.htm

Hoy, A., & Johnson, M. (2013). *Deepening community engagement in higher education.* London, England: Palgrave Macmillan.

Hofman, N. G., & Rosing, H. (2006). *Pedagogies of praxis: Course-based action research in the social sciences.* Bolton, MA: Anker.

Hofrichter, R. (2002). *Toxic struggles: The theory and practice of environmental justice.* Salt Lake City: University of Utah Press.

Israel, B. A., Schulz, A. J., Parker, E. A., & Becker, A. B. (1998). Review of community-based research: Assessing partnership approaches to improve public health. *Annual Review of Public Health, 19,* 173–202.

Jacoby, B., & Associates. (2003). *Building partnerships for service learning.* San Francisco, CA: Jossey-Bass.

Johnston, F. E., Harkavy, I., Barg, F., Gerber, D., & Rulf, J. (2004). The urban nutrition initiative: Bringing academically-based community service to the University of Pennsylvania's department of anthropology. *Michigan Journal of Community Service Learning, 10,* 100–106.

Karavoltsos, S., Sakellari, A., Antonopoulou, M., Dassenakis, M., & Scoullos, M. (2010). Evaluation of water quality in an urban park for environmental sensitization: A large scale simulation model. *Desalination and Water Treatment, 13,* 328–335.

Karlen, D. L., Andrews, S. S., Weinhold, B. J., & Doran, J. W. (2001). Soil quality: Current concepts and applications. *Advances in Agronomy, 74,* 1–40.

Karlen, D. L., Andrews, S. S., Weinhold, B. J., & Doran, J. W. (2003). Soil quality: Humankind's foundation for survival. *Journal of Soil and Water Conservation, 58,* 171–179.

Kennedy, M. (1995). Roofless women's action research mobilization. The evaluation exchange. *Harvard Family Research Center, 1,* 3–4.

Lambert-Pennington, K., Reardon, K., & Robinson, K., (2011). Creating an interdisciplinary community development assistance center: The South Memphis Revitalization Action Project. *Michigan Journal of Community Service Learning 17,* 59–70.

Lubchenco, J. (1998). Entering the century of the environment: A new social contract for science. *Science, 279,* 491–497.

Marrow, A. J. (1996). *The practical theorist: The life and work of Kurt Lewin.* New York, NY: Basic Books.

Mi, M., Stefaniak, J., & Afonso, N. (2014) Community needs assessment to reach out to an underserved population. *Medical Reference Services Quarterly, 33,* 375–390.

Mitchell, T. D. (2008). Traditional vs. critical service learning: Engaging the literature to differentiate two models. *Michigan Journal of Community Service Learning, 14,* 50–65.

Mitchell, T. D. (2015). Using a critical service learning approach to facilitate civic identity development. *Theory into Practice, 54,* 20–28.

Mohai, P., Lantz, P., Morenoff, J., House, J., & Mero, R. (2009). Racial and socioeconomic disparities in residential proximity to polluting industrial facilities: Evidence from the Americans' changing lives studies. *American Journal of Public Health, 99*(S3), S649–S656.

Morgan, R. L., Whorton, J. E., & Gunsalus, C. (2000). A comparison of short term and long term retention: Lecture combined with discussion versus cooperative learning. *Journal of Instructional Psychology. 27,* 53–58.

Morton, K., & Saltmarsh, J. (1997). Addams, Day and Dewey: The emergence of community service in American culture. *Michigan Journal of Community Service Learning, 5,* 137–149.

Newman, J. L., Dantzler, J., & Coleman, A. N. (2015). Science in action: How middle school students are changing their world through STEM service learning projects. *Theory in Practice, 54,* 47–54.

Nyden, P., Figert, A., Shibley, M., & Burrows, D. (1997). Effective models of collaboration. In P. Nyden, A. Figert, M. Shibley, & D. Durrows (Eds.), *Building community: Social science in action* (pp. 14–28). Thousand Oaks, CA: Pine Forge Press.

Pasmore, W. A., & Khalsa, G. S. (1993). The contributions of Eric Trist to the social engagement of social science. *Academy of Management Review, 18,* 546–569

Pérez-Sabater, C., Montero-Fleta, B., Pérez-Sabater, M., & Rising, B. (2011) Active learning to improve long-term knowledge retention. *Actas del XII Simposio Internacional de Comunicación Social,* 75–79.

Peters, S. (2005). Introduction and overview. In S. Peters, N. R. Jordan, M. Adamek, & T. Alter_(Eds.), *Engaging campus and community: The practice of public scholarship in the state and land-grant university system* (pp. 1–36). Dayton, OH: Kettering Foundation Press.

Pouyat, R. V., Szlavecz, K., Yesilonis, I. D., Groffman, P. M., & Schwarz, K. (2010). Chemical, physical, and biological characteristics of urban soils. In J. Aitkenhead-Peterson & A. Volder, A. (Eds.), *Urban ecosystem ecology* (pp. 119–152). Madison, WI: American Society of Agronomy.

Pouyat, R. V., Yesilonis, I. D., Russell-Anelli, J., & Neerchal, N. K. (2007). Soil chemical and physical properties that differentiate urban land-use and cover types. *Soil Science Society of America Journal, 71,* 1010–1019.

Reardon, K. M. (2000). An experiential approach to creating an effective community-university partnership: The East St. Louis action research project. *Cityscape: A Journal of Policy Development and Research, 5,* 59–74.

Reed, P., & Butler, T. (2015) Flipping the script: When service learning recipients become service learning givers. *Theory into Practice, 54,* 55–62.

Rhodes, N. J., & Davis, J. M. (2001). Using service learning to get positive reactions in the library. *Computers in Libraries, 21,* 32–35.

Rosetti, M. D., & Nembhard, H. B. (1998). Using cooperative learning to activate your simulation classroom. *Proceedings of the 30th conference on Winter Simulation.* Washington, DC, 67–76.

Rosing, H., & Hofman, N. G. (2010). Notes from the field: Service learning and the development of multidisciplinary community-based research initiatives. *Journal of Community Practice, 18,* 213–232.

Rosing H., & Odoms-Young, A. (2015). Introduction: Community-university food projects, race, and health promotion. *Journal of Prevention and Intervention in the Community, 43,*79–82.

Rosing, H., Reed, S., Ferrari, J. R., & Bothne, N. J. (2010). Understanding student complaints in the service learning pedagogy. *American Journal of Community Psychology, 46*, 472–481.

Russell, D., & Harshbarger, C. (2003). *Ground work for community-based conservation strategies for social research.* Walnut Creek, CA: Altamira Press.

Ryan, J. A., Scheckl, K. G., Berti, W. R., Brown, S. L., Casteel, S. W., Chaney, R. L.,... Mosby, D. (2004). Reducing children's risk from lead in soil. *Environmental Science and Technology, 38*, 18A–24A.

Saltmarsh, J., Hartley, M., & Clayton, P. (2009, February 13). Democratic engagement white paper. *New England Resource Center for Higher Education Publications*, Paper 45. Retrieved from http://scholarworks.umb.edu/nerche_pubs/45

Sedikides, C. (1993). Assessment, enhancement, and verification determinants of the self-evaluation process. *Journal of Personality and Social Psychology, 65*, 317–38.

See, G. K. (2010). Locally engaged education institutions: Four principles for engaging the urban environment. *Education and Urban Society, 42*, 545–571. Retrieved from http://eus.sagepub.com/content/42/5/545

Shih, T., Pittinsky, L., & Ambady, N. (1999). Stereotype susceptibility: Identity salience and shifts in quantitative performance. *Psychological Science, 10*, 80–83.

Spencer, S. J., Steele, C. M., & Quinn, D. M. (1999). Stereotype threat and women's math performance. *Journal of Experimental Social Psychology, 35*, 4–28.

Stanton, T. K., Giles, D. E., & Cruz, N. I. (1999). *Service learning: A movement's pioneers reflect on its origins, practice, and future.* San Francisco, CA: Jossey-Bass.

Steele, C. M. (1999, August). Thin ice: Stereotype threat and black college students. *The Atlantic.* Retrieved from http://www.theatlantic.com/magazine/archive/1999/08/thin-ice-stereotype-threat-and-black-college-students/304663/?single_page=true

Steele, C. M., & Aronson, J. (1995). Stereotype threat and the intellectual test performance of African Americans. *Journal of Personality and Social Psychology, 69*, 797–811. Retrieved from http://www.xavier.edu/mfi/documents/Steele_Claude_M.Stere.Thrt.inAfrican.Am.Tst.Perfm.1995.pdf

Stoecker, R., & Tryon, E. (2009). *The unheard voices: Community organizations and service learning.* Philadelphia, PA: Temple University Press.

Strand, A., Culforth, N., Stoecker, R., Marullo, S., & Donohue, P. (2003). *Community-based research and higher education: Principles and practices.* San Francisco, CA: Jossey-Bass.

Taebi, A., & Droste, R. L. (2004). Pollution loads in urban runoff and sanitary wastewater.

*Science of the Total Environment, 327*, 175–184.

Trist, E. L., Emery, F. E., Murray, H., & Trist, B. (1997). *The social engagement of social science: The socio-ecological perspective.* (Tavistock Anthology). Philadelphia: University of Pennsylvania Press.

Westbrook, R. (1991). *John Dewey and American democracy.* Ithaca, NY: Cornell University Press.

Perry, D. C., & Wiewel, W. (2008). The university, the city, and the state: Institutional entrepreneurship or instrumentality of the state. In W. Wiewel & D. C.

Perry (Eds.), *Global universities and urban development: Case studies and analysis* (pp. 3–24). Armonk, NY: ME Sharpe.

Witzling, L., Wander, M., & Phillips, E. (2010). Testing and educating on urban soil lead: A case of Chicago community gardens. *Journal of Agriculture, Food Systems, and Community Development, 1,* 67–185.

Yesilonis, I. D., Pouyat, R. V., & Neerchal, N. K. (2008). Spatial distribution of metals in soils in Baltimore, Maryland: Role of native parent material, proximity to major roads, housing age and screening guidelines. *Environmental Pollution, 156,* 723–731.

Zivi, K. D. (1997). Examining pedagogy in the service learning classroom: Reflections on integrating service learning into the curriculum. In R. M. Battistoni & W. E. Hudson (Eds.). *Experiencing citizenship: Concepts and models for service learning in political science* (pp. 49–68). Washington, DC: American Association for Higher Education.

CHAPTER 8

# ACADEMIC SERVICE LEARNING

## Community College Prenursing and Elementary Teacher Education Programs

**Anita Cuttita Ferdenzi,**
**Vazgen Shekoyan, and Sharon Ellerton**
***Queensborough Community College***

### ABSTRACT

This study examined the impact of academic service learning projects on 72 community college prenursing program biology students and preservice elementary education program physics students' confidence in learning science and professional competency skills. We examined the relationships between the project typology and student survey responses and perceptions of project benefits for providers and recipients. Data were collected using a quantitative survey, qualitative reflections, and service-learning partner feedback. Students reported increased confidence in their ability to learn and apply course content knowledge, academic efficacy, and professional competency skills. Prenursing student tutors reported a higher percentage of positive responses

*Science and Service Learning*, pages 179–219

in learning and in applying anatomy and physiology course material than prenursing students who presented information to community members in a health fair setting. The highest positive response percentages reported by preservice elementary education students, after teaching children physics in an after-school setting, were in the application of physics learning and collaborating with others as members of a team.

American community colleges play a vital role in resolving both the nursing shortage (Fulcher, 2002) and dire need for developing a highly effective teaching force (Coulter & Vandal, 2007) with science, technology, engineering, and mathematics (STEM) capable skillsets. The shortage of K–12 teachers with strong STEM skills is now considered a national crisis (President's Council of Science and Technology Advisors, 2010). The Institute of Medicine (2011) has called for the proportion of nurses with baccalaureate degrees to be increased to 80% by 2020. Community colleges are well positioned to capitalize on cultivating our nation's greatest resource to meet these mandates. They enroll more than six million students and served as the spring board for nearly half of all students graduating with a four-year degree in the 2013–2014 school year. Of all students who completed a four-year degree, 46% had been enrolled at a two-year institution at some point in the past ten years (Smith, 2015). Fewer than half of students who enter community colleges with the goal of earning a degree or certificate have attained that goal, transferred to a baccalaureate institution, or are still enrolled 6 years later (Radford, Berkner, Wheeless, & Shepherd, 2010). The 3-year graduation rate with an associate's degree is slightly more than 11%, and the 6-year graduation rate is 26%. Few finish the degree in 2 years, despite the fact that over 67% of incoming freshmen hold full-time status (Queensborough Community College, 2014, p. 9). In noting that the achievement gap in the completion of 4-year degrees is now greater than ever, retention experts Engstrom and Tinto (2008) observed,

> It is not enough to provide low-income students access to our universities and colleges and claim we are providing opportunity if we do not construct environments that effectively support their efforts to learn and succeed once access has been gained. Simply put, access without effective support is not opportunity. (p. 50)

The potential of the largest system in higher education for promoting career goals is tempered by its collective ability to deliver on this promise. Effective pedagogies in prenursing and preservice elementary teacher education program science courses are of critical importance. Research shows achievement in these courses has strong positive correlations to nursing program completion (Rodgers, 2002). The success rate among nursing students in pivotal prenursing program science courses at Queensborough Community College (QCC) demonstrates this finding. Although nearly 25% of all students entering QCC register in the prenursing program, only a small percentage of that group actually graduates with a nursing degree.

Some 4 out of 10 teachers have completed some of their math and science courses at community colleges (Shkodriani, 2004). The government's call for community colleges to help train teachers (Evelyn, 2002) who are well equipped to promote science achievement for all students will not be realized without reform in this area. In order for this new teaching force to be highly qualified and skilled in math and science instruction at both the primary and secondary levels, elementary teacher education programs must address the following three issues regarding elementary teacher preparation to teach science identified by Appleton (as cited in Lewis, Dema, & Harshbarger, 2014, p. 154). The first issue is their limited subject-matter knowledge, the second is their limited pedagogical science-content knowledge, and the third is the resultant low confidence and self-efficacy in science content and teaching. The Committee on Science Learning, Kindertarten through Eighth Grade corroborated this research finding and focused on the issue of opportunity. They attributed K–8 teachers' limited knowledge of science to their having little or no direct experience in inquiry science, limited opportunities to learn about science, lack of exposure to both the research on how children learn science, and the best practices for teaching science (Duschi, Schweingruber, & Shouse, 2007, Ch. 10). Elementary school teachers play a critical role in fostering an early interest in science in their formative students. Low achievement and negative attitudes toward science in middle and high school students have been traced back to subpar science instruction experienced during early childhood (Mullis & Jenkins, 1988).

This is a high-stakes challenge and an opportunity for community college elementary teacher education programs to work toward ameliorating the cycle of ineffective elementary teacher preparation for teaching science. Before our underprepared education students become highly qualified in building positive attitudes and strong math and science skills among their future students, they must possess these skills and attitudes themselves.

Maintaining ineffective pedagogical strategies will ensure the bleak status quo. Education historian, policy analyst, and reform advocate Diane Ravitch advises educators that the "most durable way to improve schools is to improve curriculum and instruction" (Ravitch, 2010, p.). This research study sought to do just that through its investigation of the effects Academic Service Learning pedagogy had on the prenursing and preservice elementary students' positive responses to items regarding their learning and application of physics and biology material.

## LITERATURE REVIEW

### Underprepared Prenursing and Preservice Elementary Education Students

Professional development programs in teaching and nursing share an imminent social purpose in promoting the health and well-being of our society. The utility of learning required science courses concepts reaches far beyond the extrinsic goals of passing exams required for degree attainment. For prenursing students, the mastery of anatomy and physiology is essential for competencies in surgical medical procedures and treatments (Nyatangam, 2005). "The possession of skills to improve client health which require higher level thinking is essential for the safe and effective care of clients in today's health care environment" (Kostovich, Poradzisz, Wood & O'Brien, 2007, p. 26). Community college nursing students without an adequate science background are at a major disadvantage as they undertake important gateway biology courses, such as anatomy and physiology. Perhaps the largest stumbling block is the nearly 50% attrition rate in this high-stakes gateway course into the nursing program. Students are underprepared with both inadequate science backgrounds and the necessary study skills needed to navigate this fast-paced, detail-oriented course. Often, by the time the students taking the course understand how to study and the necessary time commitments, it is already too late because they have failed one or more exams. An introductory human biology course recommended to underprepared prenursing students is BY-110. The purpose of this preliminary course is to introduce prenursing students to relevant biological concepts and prepare students for the rigor of the courses they will be taking. Even in this preliminary course, unfortunately, many students do poorly, are discouraged, and change their major, leaving behind their dreams of a career in health.

Biology is a discipline that is multitiered—studied at the cellular as well as organismic level, and at the anatomic as well as the physiologic level; with all levels interrelating and impacting each other. When the science educational system is working well, a student progresses through elementary science courses throughout years of schooling, so the fundamental concepts from chemistry and physics, along with the multiple exposures to increasingly complex biology courses, build on one another creating a holistic understanding of biological concepts, best preparing a student to take college biology courses. This is not the reality for many students at the community college. For some, biology is a long-forgotten course as they come to retrain for a new profession. For others, biology has not been part of their studies since early childhood, as occurs in vocational pathways schooling in some countries. For these students, the myriad biological details they must learn remain isolated facts to memorize rather than interweaving threads to integrate.

The Trends in International Mathematics and Science (TIMSS) 2008 study research revealed that the United States is losing ground in a very competitive market (Fathe & Kasabian, 2008). Teachers' beliefs about whether or not they have the ability to teach elementary science is fundamental to their students' learning (Plourde, 2002) and will impact their students' attitudes toward learning science (Krajcik, Czerniak, & Berger, 2003). For education students, mastery of science content is essential for a positive trajectory toward becoming an effective science teacher. We know that students' beliefs play a pivotal role in their academic success. Self-efficacy beliefs are content specific and based upon their perceived success or failure at a given task. The cycle of impoverished math and science skills and efficacy, well established in the literature, will persist with traditional pedagogy. Students in these programs need authentic activities designed to facilitate learning and focus students on meeting career and transfer goals.

Research shows teachers' attitude and instructional strategies have a substantial influence on the student-learning outcomes in science (Shrigley, 1983). Teacher-preparation researchers, however, warn that many programs for preservice elementary teachers do not provide the kinds of learning experiences suggested by new developments in the science of learning. Teacher preparation courses in methods of teaching science emphasize technical skills rather than decision making, theory, and reasoning. Teachers who were taught science in a didactic way tend to teach the way they were taught. Teacher education researchers recommend that teachers become comfortable in teaching science by exploring the process-oriented-inquiry method activities intended for the children they teach (Martin, Jean-Sigur, & Schmidt, 2005).

The National Science Education Standards (National Research Center, 1996) advised that, in order for science education reform to be accomplished, professionals must be engaged in active learning that builds their knowledge, understanding, and ability. The Standards warn this new vision for teaching science is nearly impossible to convey to students in schools if the teachers themselves have never experienced it and that there be a substantive change in how science is taught that is equal to the substantive change in professional development practices (p. 56). This is especially critical for elementary school teachers with general certification whose only science courses taken are at the introductory undergraduate science level.

## BEST PRACTICES IN SCIENCE EDUCATION

### Physics

Traditional lecture-style physics courses are not effective in improving students' conceptual understanding of physics. Active engagement techniques help students gain deeper understanding of the concepts (Hake, 1998; Iverson, Briggs, Ruiz-Primo, Talbot, & Shepard, 2009). The common strand running through different nontraditional instructional strategies in physics is the focus on opportunities for students to improve conceptual understanding of physics problems and reasoning ability in experiential and social contexts. Although the very structure of large college lecture halls precludes opportunities for deep learning experiences, recitation classes with smaller student-teacher ratios impede problem-solving processes when students are passive witnesses to teacher-centered instruction. Students learn more physics in classes where they interact with faculty, collaborate with peers on interesting tasks, and are actively engaged with the material they are learning (Knight, 2002). Student-centered learning environments promote efficacy-building activities through challenge and support through explicit expectations and multiple learning tasks.

In a study of the innovative studio model employed in the Student-Centered-Active-Learning-Environment Undergraduate Programs (SCALE UP) Project, researchers (Beichner et al., 2007) found that the most effective aspect of the instructional approach was in the tight alignment between the course learning outcomes and labs, lectures, and recitations. The benefit of combining discrete components of the course into a single studio-style class were further evidenced in its application in multiple sites. The students in SCALE-UP classes gained a better conceptual understanding than did their peers in traditional lecture-based classes. In the first and second semesters of introductory physics, students performed better on a variety of conceptual surveys. Faculty involvement with students and active, self-directed learning contributed more than anything else to measurable student success (Astin, 1993).

## Biology

Biology faculty continually debate how to best serve these students. Many agree that traditional lecture methodology may not be the best vehicle to support concept integration for these students, which is supported by pedagogical research. Research within biological sciences in higher education has shown that students perform better when active learning techniques are incorporated into the classroom. These methodologies allow students opportunities to construct and integrate concepts (Michael, 2006). Wilke and Straits (2001) found students who learned topics through "discovery learning" performed better than those taking traditional coursework. Lumpe and Staver (1995) demonstrated that collaborative groups learned biological principles better than when students studied alone. Constructivist methodology was used by Burrowes (2003) to teach biology while also teaching a traditional lecture section as a control group. The experimental group performed significantly better on exams than did the control group, particularly excelling in "scientific thinking"-type questions. Using a variety of active learning techniques, including clickers, Freeman et al. (2007) also found that active techniques improved student academic performance.

A further issue is that students are underprepared in biology itself. Many biological principles are based in chemistry and physics; without the benefit of these foundational science courses, the large amount of information thrown at the student is often difficult to integrate as the course rapidly progresses. As an example, Hopp (2009) found that students entering an anatomy and physiology 1 (A&P1) class performed significantly better when having taken a chemistry course in advance. Many of the students enter A&P1without foundational coursework, leading to the large dropout and failure rates we see nationwide (Harris, Hannum, & Gupta, 2004). This issue is exacerbated as many colleges across the country have no science prerequisites for A&P1. The only prerequisite at CUNY Queensborough Community College, for example, is that the student has passed through reading remediation. At this college, no other required prerequisites have been allowed due to the tightly proscribed prenursing program credit structure. There are no additional credits available to force requirements for A&P1, to the detriment of the large numbers of students who begin A&P1 hoping to become a health professional only to have those dreams quickly fall to the wayside. Students are encouraged, therefore, to take a pre-A&P course to provide this much-needed support. Many students to do not choose this option.

Research on the best practices for promoting college student retention indicates that high-impact practices, such as collaborative learning, student research, writing-intensive classes, learning communities, global learning, and common intellectual experiences, provide powerful contexts that engage students in the type of deep and meaningful learning that promotes greater academic success rates. These educationally purposeful activities have been shown to benefit historically underserved students more than traditional students (Kuh, 2008). A recent study of service learning's influence on career and technical education students' retention at Queensborough, revealed that service learners were more likely to be retained than nonservice learners. Of students who enrolled in fall 2012 and participated in service learning, 80.7% were retained in spring 2013 as compared to 73.5% for nonservice learners (Ellerton et al., 2014). Academic service learning pedagogy supports the principles of professional learning through activity, reflection, collaboration, passion, and community (Shulman, 2004) and offers unique opportunities to learn science.

## THE CATALYTIC HIGH-IMPACT PRACTICE: SERVICE LEARNING

Kuh's (2008) seminal analysis of the large dataset collected through the National Survey of Student Engagement, identified educational practices that correlated with positive educational outcomes of college students and described them as (a first-year experiences; global/world learning; (b) learning communities; (c) collaborative projects; (d) student research; writing intensive courses; (e) internships; (f) capstone courses and projects; and (g) service learning. These strategies are referred to as high-impact practices (HIPs). The study further reported that HIPs benefited all students. Data also revealed that "historically underserved students tended to benefit more" (p. 17) and showed higher gains in first/second year retention rates and grade-point averages. This finding is especially important in view of the fact that underserved students are often not as academically prepared as others for higher education and more likely to drop out. The observations of Hanelle Niemi (2009), a leading Finnish expert in teaching and learning, offers this insight: "Young people who fail within school, can and often do develop through unconventional activities that provide the 'experience and enjoyment of learning'" (p. 6).

Among these unconventional instructional strategies, service learning is the HIP whose catalytic properties have the greatest potential for promoting student growth beyond the bounds of intended and anticipated outcomes (Honnet & Poulsen, 1998). Perhaps service learning inspires the "flow and exhilaration in learning currently experienced outside formal learning contexts rather than within them" (Niemi, 2009, p. 6). Berzelius identified catalytic power as the ability of substances to awaken affinities which are asleep at particular temperatures by their mere presence, and not by their own affinity (as cited in Roberts, 2000, p. 1). The "affinities" that academic service learning projects awaken in the learner's meaning-making processes are reviewed in this section and organized in terms of the high-impact practice's catalytic properties.

### Catalytic Property 1: Constructivist Pedagogy

Service learning provides educators with a rich social core for powerful learning activities. Eyler and Giles (1999) reported that students said they were more motivated to work harder in classes, experienced deeper understanding of course content, and were better able to apply subject matter to real-world problems after service learning activities. In another study of community college students, investigators found students who completed one or more service learning courses reported significantly higher learning outcomes than nonservice learning students across multiple courses, concluding that service learning provided the conditions for learning (Prentice & Robinson, 2010).

Embedding learning objectives in real-world contexts is grounded in the philosophy of John Dewey. His interest in community and democracy inspired his belief that learning occurs through the experience of solving relevant problems, through interaction with others, and reflecting on the experience, which is the very essence of service learning pedagogy. Piaget extended the work of this leader in progressive education with his cognitive theory of development, where learning is grounded in the interface between mind and environment (Lutz & Huitt, 2004). Piaget viewed novel experiences in one's environment as disruptive to the equilibrium between one's current understanding. As such, disequilibrium is the impetus for intrinsic motivation. Experiences that do not fit into our current understanding of the world are problematic, triggering a need for further inquiry through rigorous experimentation and reflective thinking, which Dewey believed "impels to inquiry" (1933, p. 7). He further believed that, although we learn by doing, experiences are not educative if they are not reflected upon in a rigorous manner.

In service learning projects, reflection connects the course concept to the experience facilitating connections between practical applications of knowledge. In contrast to the passive processing of information for the purpose of reproducing the materials to be learned for academic assessment, or surface knowledge, students who comprehend the meaning of the material to be learned (Richardson, 1994, p. 463) are intrinsically motivated and engage in reflective knowledge construction derived through active learning. The concepts of "tacit knowledge," "knowing in action," and "reflection in action" were studied by Schon (1987) in his research on reflective practitioners. He identified knowledge in action as the knowledge of practice developed by experienced professionals and evidenced in the skills exhibited through spontaneous application in performance. Although this form of knowledge is tacit in its automaticity, the process of observation and reflection allow the practitioner to capture their dynamic competencies. The power of the reflective cycle is attributed by Rodgers (2002) to its ability to slow down teachers' thinking to what is and then shift the weight of that thinking from their own teaching to their students' learning. Reflection in action requires the process of formative evaluation and "thinking on ones' feet." (Schon,1983, p. 40).

Academic service learning projects provide powerful contexts engaging preservice community college nursing and education students in deep learning experiences and opportunities to practice the reflective-thinking habit. It offers students the opportunity to connect course theory with experiences and to build links between theory and practical applications. In a study of nursing students, researchers found that through this practice, nursing students were better prepared to function professionally in the community setting and were empowered in their role of nurse as teacher of preventative health care (White, 1999). This finding is further supported in another study which suggested that the service learning teaching project is a way to promote both student learning and professional growth, producing strong evidence for optimizing academic success by helping to teach others (Astin & Sax, 1998).

Research on service learning teaching projects in the sciences has yielded similar positive results. In a study of a service learning project to teach plant genetics to secondary students, postproject surveys showed positive effects on undergraduates' content knowledge for both science and nonscience majors participants (Chrispeels et al., 2014). Another study of a service learning project involving undergraduates in a nonmajors biology course teaching high school students suggested that learning outcomes benefited both providers and recipients (Hark, 2008). A study of the implementation of community-based learning in Ireland found a significant improvement in student engagement and confidence in the course after preparing and presenting posters and demonstrating experiments to secondary school students during Science Week.

The federally funded Learn and Serve America Higher Education STEM grants initiative supported the development of service learning programs throughout the country in individual colleges and multistate consortiums. In a report highlighting the results of over 1,000 projects developed and implemented over a 3-year period, research on students' responses to the pedagogy showed an expressed appreciation for the experience of applied, problem-based, real-world learning. The research conducted by the Midwest Compact Consortium of Colleges and Universities involved in the initiative showed that approximately 75% of the participants agreed that service learning enhanced their understanding of scientific concepts in courses and increased their interest in STEM (Hamlernick, 2013). The projects in this reform initiative formed two categories: STEM aspiration in K–12 students and course development through higher education and issue-based explorations and applied work in STEM disciplines related to environmental stability, food sustainability, and health. The diversity among the STEM topics, purpose, community partners, course structures, target learning outcomes and individuals involved, contribute to challenges in designing a uniform assessment methodology for the pedagogy.

## Catalytic Property 2: Diverse Nature of Service Learning

Service project designs vary in terms of the relationship between the service and the learning. Sigmon (1997) differentiated relationships in service learning using various combinations of upper- and lowercase fonts for visual emphasis: (a) service learning–learning goals are primary and service activities are secondary; (b) Service learning–service outcomes are of primary importance to learning goals; (c) Service Learning–service and learning goals are completely separate; and (d) SERVICE LEARNING–service and learning goals are balanced and beneficial to all participants (p. 2).

The need for a definitive characterization of service learning was identified by Furco (1996) and elaborated on Sigmon's work on SERVICE LEARNING criteria to guide his examination of the different levels of reciprocity inherent in various types of service programs. He studied experiential service programs such as internships, community service, volunteerism, educational fieldwork, and service learning to discern the level of balance between their primary intended purpose and focus to derive a more precise interpretation of their differences. Furco offered a detailed definition for each type of service program and conceptualized a model to illustrate the degree of emphasis each program has on the *Beneficiary* (recipient-provider) and *Focus* (service learning). The terms *Beneficiary* and *Focus* are located at the central point of the top of the model with the terms *Recipient* and *Provider,* and *Service* and *Learning* situated respectively, on a continuum on either side of the terms. Each service is depicted as a block in a graduated pyramid with Service Learning located in a central position at the top of the pyramid. Volunteerism and community service are in hierarchical positions depicting their emphasis on the "recipient benefit and focus on service" on the left side of the continuum and field education and internship are in hierarchical positions located at the opposite end of the model's continuum, illustrating their emphasis on "provider benefit and focus on learning." The model clearly depicts Service Learning as the service having the most balance between service and learning.

## Catalytic Property 3: Interpersonal Dynamics Between Service Learning Partners

Research conducted by Georgetown University scholars (Bass & Elmendorf, 2012) offers further insight into enhancing the service learning experience for the provider and recipient through the lens of social pedagogies.

The researchers operationalize social pedagogies as follows: design approaches to teaching and learning that engage students in authentic tasks that are communication intensive, where the representation of knowledge for an authentic audience is absolutely central to the construction of knowledge in a course. All service programs provide a social core, real-world context for the learner. However, the service learning project provides the learner with the unique opportunity to represent the knowledge they have gained to an authentic audience for whom the information matters. This personal connection is key to maintaining the balance between the learner and the recipient and taps into the affective domain. The objectives of this research study were to examine the impact that service learning projects, which were developed for required gateway science courses in the Queensborough Community College prenursing and preservice elementary education programs, had on the students' perceived ability to (a) learn material in required science courses more effectively, (b) apply course content during service learning experiences, (c) have more confidence in self-regulated learning skills, and (d) have more confidence in professional competencies. This study sought to examine the relationships between the percentages of positive responses and the type of project in terms of course structure, project activities, and level of interaction with service learning partners to inform service learning practice to promote academic success.

## DESIGN

A quantitative and qualitative approach was adopted. A postsurvey was administered to students in the nursing and education programs. Survey questions asked students' views on their attainment of skills and knowledge based on a 5-point Likert scale. Qualitative data and open-ended responses from the students were used to gain descriptive information for each group. Students completed postproject reflections containing short-answer questions on their ability to learn course material more effectively, connect course material to service learning activity, and enhance confidence in workplace and self-regulated learning skills. The survey included an open-ended question in which students were asked to describe a specific project experience.

Data for this study were analyzed for students enrolled in the (a) elementary preservice teacher education learning community–required Introduction to Physics course during the Fall 2013 and Spring 2015 semesters and students enrolled in the (b) prenursing program–required Biology courses Anatomy & Physiology 2 (A&P2, BI-302, as well as the (c) prenursing nonrequired pre A&P course for students with weak backgrounds in science (BI-110), during the Fall 2013 and Spring 2014 semesters. The data from these semesters were analyzed within course cohorts. The number of prenursing students responding to the survey in BI-302 for Fall 2013 and Spring 2015 were 12 and 16, respectively. The number of prenursing students responding to the survey in BI-110 for Fall 2013 and Spring 2014 were 10 and 7, respectively. Of the 28 students enrolled in BI-110, 82% were female, 12% were male, and 6% were nondisclosed. The demographic distribution of these students was 12% African American, 41% Hispanic, 12% Asian, 24% White, 12% from other races, and 6% nondisclosed. The demographic distribution for BI-302 was 93% female, 7% male, 43% African American, 21% Hispanic, 25% Asian, 11% White, and 4% identifying themselves with more than one race. The number of education students enrolled in Physics for Fall 2013 and Spring 2015 were 11 and 16, respectively. Out of those 27 students, 24 were females. The demographic distribution of those students was 30% White, 30% Hispanic, 19% Asian, 11% African American, and 12% from other races or identifying themselves with more than one race.

Students ranked the impact that the service learning experience had on their learning and applying science course content and confidence in their self-regulated skills, among other items. The survey consisted of 22 questions with graded responses on a 5-point Likert scale, as well as open-ended questions asking students to provide an example of their experience during the service learning project. Questions pertaining to professional work-skill acquisition such as team membership; expressing themselves through writing and speaking; and self-regulated learning such as solving challenging problems; follow directions completely; ask questions in class; and complete projects or assignments on time. Possible student responses included "Not at all," "A little bit," "Somewhat," "A lot," or "Not applicable." The possible student responses to the items directly related to learning and applying science content during the service learning experience and to civic growth included "Strongly disagree," "Disagree," "Neutral," "Agree," and "Strongly agree."

## IMPLEMENTATION

The Implementation section is organized by discipline. Course descriptions for Physics, Biology, and Education courses involved in the academic service learning projects include commentary on the constructivist components in each. The Physics course section, which includes a detailed description of the Inquiry lesson on magnetism modeled by the Physics professor, is provided to illustrate the Education course content embedded in the service learning project. Detailed descriptions of the implementation of each service learning project are located after each course description.

## Description of Science Courses and Contexts for Academic Service Learning Projects

Academic service learning (ASL) is one of seven high-impact practices (HIPs) supported by the Queensborough Community College Academies. Faculty design projects to take student work out of the classroom and into the community to facilitate learning and promote civic engagement. In terms of structure, the ASL projects in anatomy and physiology described in this study are within the context of a science course. In the pre-A&P BI-110 course, the students were provided an extra credit project option to develop an informational poster on a particular disease for a health fair. In A&P2 class conducted as a traditional course, an extra credit service learning option to peer tutor provided students an opportunity to engage in constructivist learning.

The Physics ASL project was embedded in the context of a learning community. Learning communities are another High Impact Practice offered by the college. Learning communities are composed of a cohort of students taking two or more courses taught by two professors, linked by a theme, and designed to assist students in seeing connections between disciplines to enhance faculty and student engagement. The Making Waves in Physics learning community is especially designed for preservice teaching students who are enrolled in both a required introductory course in Physics (PH-101) and Education (EDU-101). The content in each course is articulated to promote the achievement and self-efficacy of future teachers in science through constructivist learning experiences.

## Contemporary Education: Principles and Practices Course Description

EDU-101 is a required course in the elementary education program. It is a one semester long, four credit hour course examining the historical, philosophical, and sociological foundations of American education. Students study the evolution of educational theory and research-based practices that promote social, emotional, and cognitive development and enhance learning. Attention is paid to comparative analysis of past and contemporary political, philosophical, and sociological factors that have influenced and shaped educational decision-making. The course introduces students to the reflective decision-making model through readings, in-class activities, and discussions. Students are actively involved in their own learning process. This class employs high-impact learning opportunities such as Academic Service Learning, Writing Intensive, Collaborative Learning through ePortfolio, and Learning Community design. Students have the opportunity to participate in a culminating academic service learning activity entitled Physics Friday, held at the QCC After School Program.

## Principles of Physics Course Description

The PH-101 is a required course in the elementary education program. It is an introductory conceptual physics course with no mathematics prerequisites. It is a one semester long, four credit hour course (three hours lecture and two hours laboratory per week). It satisfies a science course with a laboratory requirement for Liberal Arts and Sciences majors. It is also one of the science courses that elementary education majors are required to take. Typically, one third of PH-101 course population is elementary education majors. It is a multisection course taught by multiple instructors. There are about 25–30 students in an average lecture class and not more than 24 students in the lab section. The course presents a modern overview of the world around us, from the subminiature world inside an atomic nucleus to the vastness of outer space. The conceptual understanding of physics is the major goal of the course. By the end of the semester, students are expected to be able to describe the concepts associated with topics in physics and use them to write answers to conceptual questions as well as to solve simple physics problems (the amount of mathematics involved is minimal). Some of the topics covered in the course are linear motion, forces and Newton's laws, conservation laws, and selected topics related to rotational motion, heat, vibrations, waves, electricity, magnetism, and optics.

## Academic Service Learning Project in the Education Physics Learning Community: Physics Friday

In this model, the academic service learning project is embedded within the Making Waves in Physics Education learning community. The Making Waves in Physics Education learning community format is designed to provide future teachers with deep learning experiences in the effective teaching and learning of science content from two disciplinary vantage points. Students learn how to use a variety of learning and teaching strategies from Education 101 as they flex their perspectives from that of a current student learning Physics 101 content material to that of future teacher, studying underlying pedagogical principles for effective science instruction. Students further apply course content in a common academic service learning project providing the opportunity for experiential learning. In this project, students deliver lessons on physics topics to elementary school children in a community after school center. Both courses actively involve students in their own learning process. Other models for physics "after-school" and "during-school" elementary school–level service projects embed the project in stand-alone introductory physics courses (Othmer & Sealfon, 2010).

At the beginning of the second half of the semester, the PH-101 instructor assigned course physics topics to groups of students. Students were allowed to exchange topics with each other during the first week. All PH-101 students had to prepare a plan to teach their assigned physics topic to elementary school–aged students. Students conducted research on their topic and shared ideas with classmates as they collaborated to perfect a variety of inquiry activities to teach the school children to experience and learn the physics topic in their course. Two revisions of the lesson plan were required after receiving extensive feedback on science content from their physics professor in PH-101 and on instructional strategy content feedback from their education professor in EDU-101. The lesson plans were evaluated based on the following criteria: (a) correctness and relevance of the physics content and planned activities in the lesson plan; (b) following constructivist approach in lesson plans, with special emphasis on discovery activities; (c) logical organization of the lesson plan and clarity of writing. After the final versions of lesson plans were accepted by the PH-101 instructor, one or two Fridays (depending on the number of participating preservice teaching students) were chosen for students to visit the afterschool academy to teach their physics topics to the children. The event is known as Physics Friday.

The college academic service learning office arranged for transportation to and from the college to the Queensborough After School Academy. The after school academy staff facilitated the event by providing a separate room and sign for each physics topic. The after school children were divided into four groups of five mixed-age children. The groups rotated between the "Buoyancy, Friction, Magnetism and Inertia" rooms, spending about 20–30 minutes in each room. Each group was accompanied by the after school academy staff member. The PH-101 and EDU-101 instructors visited each group periodically to provide feedback and formative suggestions between short intermissions as groups transitioned between rooms. Preservice elementary education students took turns in leading the inquiry learning activities in each rotation. Students were required to submit pre- and postreflection assignments on service learning to the ED-101 instructor (Table 8.1).

The topics chosen for lesson plans by the physics instructor are listed in the left column of Table 8.2. The central column of the table describes general ideas/concepts that can be included in the topic. Students were encouraged to find materials to be used in their experiments. In addition to that, the instructor notified students what equipment and materials he could provide from the college laboratory equipment directory (a few supplies were funded by the college's service learning office). Students bought

**TABLE 8.1 Pre- and Post-ASL Physics Project Reflection Prompts**

| Preservice Learning Reflection Prompts | Postservice Learning Reflection Prompts |
|---|---|
| 1. What do you expect to learn through this service learning project?<br>2. What types of experiences do you hope your participation in this project will provide you with?<br>3. Describe the activities you anticipate you will share with your ED 101 partners.<br>4. Do you imagine that this service learning experience can assist you in understanding ED 101 course concepts?<br>5. How can this service learning experience impact your professional development as a future teacher?<br>6. Do you think your participation in this service learning project can promote insights into your own learning process? | 1. Describe 1–5 central course concepts or theories you learned in your course that relate to your service learning project.<br>2. How did you observe/use/apply these course concepts or theories in your project?<br>3. Explain how your activities deepened your understanding of these course concepts or theories.<br>4. Did your activities have an impact? How do you know? |

additional inexpensive supplies (such as ropes, eggs, paper clips, sandpaper, etc.) to be used in their inquiry science activities.

## Constructivist Approach in Education Course

In the Education course, the constructivist approach began with a baseline philosophy of education statement and reflection on past experiences in learning science and related efficacy and attitudes toward learning science. Students elaborated on this reflection to describe the instructional strategies they experienced in learning science in terms of teacher-centered strategies such as direct instruction, drill and practice, lecture; dialogue oriented- question and answer, discussion; modeling and student-centered discovery learning or inquiry learning. Students were then asked to write a lesson plan on the states of matter and then later analyze their plan to determine the instructional strategies they employed in their lesson. Students conducted research on best practices in science education and were then directed to adapt their original plan into a constructivist inquiry lesson plan format. Students were then directed to apply the next course topic to their lessons and write a commentary on how and why they have incorporated activities to meet the diverse needs of learners, to identify the science process skills the children will use during the lesson, and to identify which level of Blooms Taxonomy of Knowledge their plans will engage. Upon critical review of their baseline lesson plan, students compared their original lesson

**TABLE 8.2 Physics Friday ASL Project Alignment to Physics Course Learning Objectives**

| Physics Course Topics | Concepts and Ideas | Equipment |
|---|---|---|
| Inertia of objects at rest | Objects at rest tend to stay at rest; more massive objects resist more and get less acceleration | Household items |
| Inertia of moving objects | Moving objects tend to keep their directions and rates of motion same. More massive and faster moving objects resist to changes in motion more | Roller skates, hover pucks |
| Friction force | Static vs. sliding friction. Its independence of surface area and speed; its dependence on surface quality and normal force | Household items |
| Free fall and drag force | Objects in free fall with same acceleration. Drag forces slowing down the fall. Discovering the dependence of drag force on surface area and speed. | Balloons |
| Newton's third law | Identifying action-reaction forces. Discovering that action-reaction forces have equal strengths. | Spring scales, balloons, balloon helicopter |
| Buoyancy | Properties of buoyant force; understanding terms such as density, displaced volume, conditions of floating | Buckets, beakers, graduated cylinders |
| Magnetism | North and south poles, magnetization and demagnetization, how compasses work. | Magnets, compasses, audio tape |
| Optics | Reflection and refraction, total internal reflection, dispersion, colors, color addition and subtraction | Lenses, prisms, primary color light sticks, color filters, handheld spectroscope |

plans to their revised lesson plans, which incorporated best practices in science education. Students were asked to evaluate whether their plans reflected a didactic teacher-centered approach or a student-centered discovery and inquiry approach.

Students' construction of their understanding of the inquiry approach to teaching science was further developed in the physics class component of their learning community. Professor Shekoyan deliberately modeled the inquiry approach in his delivery of a lesson on magnetism. Students were asked to describe how Professor Shekoyan captured their attention and to identify the question, challenge, or interesting event that piqued their curiosity. The students were then directed to provide a description of the exploration phase of the lesson. They discussed the materials they manipulated to explore and gather information about magnetism. They reflected on how they experienced learning through their own actions and reactions, and how they were introduced to aspects of a new idea, concept, or thinking skill. The students were then asked to reflect on the shared activity that provided opportunity for communication through the exploration phase. Next, students described the concept-development phase of the Inquiry Model lesson plan. Students reflected on the direct teaching methods their physics professor used, how the information was organized, and the new terminology they learned. They described the similar experience in a related activity that allowed them to apply the definitions, explanations, terms, and concepts they had learned. Finally, the students reflected on the similar experiences they had in a related activity that allowed them to apply the definitions, explanations, terms, and concepts they had learned. This metacognitive exercise was designed to anchor the student's direct experience in learning, level their diverse experiential backgrounds in science with a common and shared activity, and serve as a model for designing their academic service learning lesson plans. Students were required to submit pre- and postreflection assignments on the ASL experience to the ED-101 instructor (see Table 8.1).

## Constructivist Approach in Physics Class Demonstration

In general, a constructivist approach to learning is guided by the following four principles: learners construct their own meaning; new learning builds on prior knowledge; learning is enhanced by social interactions; and learning develops through "authentic" tasks (Good & Brophy, 1994). Constructivist learning implies that the concept follow the action and not the other way around (Cooperstein & Kocevar-Weidinger, 2004).

In teaching PH-101, the instructor employed a constructivist approach. Students engaged in discovery/exploratory activities that could lead to the discovery of a concept or principle (or at least to some aspect of the concept) rather than introduce topics through direct instruction. While contemplating the meaning of information gained through the activity, students' thinking was triggered to make connections between new knowledge and prior knowledge and reflect on contradictions that might exist between the two. Typically in this phase, students worked together in small collaborative groups (two to four students per group, depending on the classroom setup). The group members discussed their ideas and conclusions with fellow group members and then recorded their own answers. The instructor collected students' answers and led the class discussion. After the input from the instructor, students continued on to engage in other activities to test or further strengthen their newly constructed knowledge through authentic tasks. The instructor then summarized their findings and provided direct instruction on the concepts related to student discoveries derived through direct investigation. The instructor gave students a homework assignment (in most of the cases similar to the homework assignments assigned in the traditional format) and later clarified course concepts as students discussed their answers.

Rather than being told that any two interacting objects exert forces of equal strength on each other (Newton's third law), students, for example, were engaged in a discovery activity that led them to the conclusion that at least for the case of pulling objects, the two forces are of equal strength (the instructor actually handed out spring scales to students so that they could pull each other and directly measure the strength of the pulling forces). During discussion, the instructor suggested the possibility that the forces would be equal in all interactions, not just for pulling forces. After that, students discussed the possibility of testing the suggested hypotheses for other experiments. The instructor provided a formal lecture on Newton's third law only after students had experienced these activities.

## Overview of Academic Service Learning Project in Prenursing Sequence Biology Courses

Academic service learning was introduced into BI-110 classes to involve students in projects to connect what they were learning in the course to the real world and excite them to continue in a health career as they learned more about the subject matter. The project chosen was that students would present posters on diseases at the health fair. The rationale was that poster presentations would force students to negotiate the language of science, provide relevance to their studies, and allow students to teach others what they learned while using the language of science. In addition, ASL was performed during two consecutive semesters in an A&P2 course, where A&P2 students tutored and mentored peers in A&P1 and A&P2 who were not performing well in class. This project served a double purpose in that it forced A&P2 students to review and better learn their own course material as well as the foundational information found in A&P1, while providing much needed assistance to their poorly performing peers.

In the pre-A&P BI-110 course, the students were provided an extra credit project option to develop a poster for a health fair. Students were divided up into five groups, with each student tasked to research an aspect of cancer. Revisions of student work occurred across the duration of the project to ensure accuracy of information and understanding of content. This work culminated with the presentation of their work to visiting student groups.

The service learning occurring in the A&P2 course was also offered as an extra credit option. Tutoring assignments were designated after the first examination. Stronger students were assigned to tutor peers in their own classroom who did not perform well on the exam. Depending on their academic skill, the weaker students were either assigned to tutor weaker A&P1 students, or pre-A&P (BI-110) students. In this way, students were forced to review A&P to properly prepare for their tutoring. Tutoring continued throughout the semester. Both tutors and tutees were responsible for completing a reflective evaluation form that provided the details of each tutoring session as well as questions related to the success of the session (for the tutor reporting form, see the Appendix). At the end of the semester, the service learning students in both the pre-A&P course and the A&P2 course were asked to complete an IRB-approved survey, which allowed students to report on their perception of their own learning and their confidence level in a variety of skills important to success at school and in the workplace.

***Course Description of BI-110: Fundamentals of Life Science***

Fundamentals of Life Science (BI-110) is a multisection introductory biology course for nonmajors with no laboratory component. A few sections of this course are designated specifically for underprepared prenursing students and for students in the Medical Office Assistant Program (MOA), and is taught as a one-semester human biology course. This course is required for the MOA students but is not a required course for prenursing students. Despite that, BI-110 is highly recommended for all prenursing students without a strong science background. Typically, there are 25 to 30 students in a class, with approximately three quarters of the students planning to pursue a career in nursing. The purpose of this preliminary course is to introduce these students to relevant biological concepts and prepare them for the rigor of the courses they will be taking. This course presents a survey of basic human anatomy and physiology, covering all aspects of human biology from the atomic level, building to molecules, cells, tissues, organs and organ systems. By the end of the semester, students are expected to know the vocabulary of human anatomy and be able to recognize and decode scientific and anatomical terms. They are also expected to be able to identify anatomical structures and understand basic principles of how organ systems function.

Traditionally, BI-110 is a three hours/week lecture course. Students are given reading assignments from their textbook which they are expected to read prior to the lecture, though there is no mechanism in place to monitor students coming prepared to class. Many students report that they find reading the book confusing and prefer to listen to lecture to initially learn the concepts. Periodic postlecture homework is assigned, graded, and returned to the students. Also periodically, worksheets are distributed during class time, and groups work together to complete these forms. In the traditional delivery of this course, concept introduction precedes homework and group assignments.

## Academic Service Learning Project in Pre-Anatomy and Physiology: Health Fair

Academic service learning is a constructivist approach to teaching pre-anatomy and physiology. Early in the semester, groups were formed that worked together for the purposes of developing a poster to present at either a health fair or a science fair. Students were given scheduled assignments to help them prepare for the fair and research the information they needed to assemble the poster. The students selected their choice of particular topic, but all were based on human disease, specifically different forms of cancer. The project required the students to investigate the impact of the specific cancer and its impact on normal body structure and function. Students began to research this topic long before they learned the body systems in class. Over the course of the semester, concepts being taught in class supported different aspects of their research, and they formed strong connections to the concepts as they realized that what they learned in class had important, real-life applications. As they learned concepts they had already struggled with in their research, many misconceptions were clarified.

In the fall of each semester, the Biology Club at the college hosts an "I Love Science Day" fair, while in the spring of each semester the college health office hosts a health fair. For both events, a variety of attendees arrive with varying backgrounds in science. High school students, college students, and the community at-large are invited, and the fairs are well attended. In particular, high school student groups travel from poster to poster to learn about the given topic and ask questions of the presenter.

For both the fall and spring events, the BI-110 students had the same extra credit project, which was to prepare a poster for presentation that provided information about a disease process and its impact on human anatomy and physiology. Students were provided an extra credit project option to develop a poster for a health/science fair. Students were divided up into five groups; each assigned a different form of cancer. Each of the five members of a group was assigned a different aspect to prepare, and each was responsible for their own bibliography.

Since the fairs take place during college club hours, most of the students are able to stay with their poster and speak with visitors that arrive and ask questions. Students distribute their handouts along with giveaways they may have successfully obtained from health organizations that deal with the particular disease they have chosen, such as the American Cancer Society.

***Course Description: BI-302: Anatomy & Physiology***

Biology 302, lecture course: Anatomy & Physiology 2 (BI-302) is the second semester of a year-long Anatomy & Physiology course. This two-semester sequence is a requirement for the nursing program and certain other health-related professions. Each semester course is four credits and has a laboratory component. The class lecture meets three hours/week and has approximately 30 students, and there is also a three-hour laboratory component that generally complements the learning done in lecture. The lectures and labs are often taught by different instructors. This course presents a detailed review of human anatomy and physiology, covering details of human biology from the atomic level, building to molecules, cells, tissues, organs, and organ systems. By the end of the two-semester sequence, students are expected to have a sophisticated understanding of the human body. Traditionally, BI-302 lectures involve three classroom hours per week. Students are given reading assignments from their textbook, though there is no monitoring of this process. Students are only assessed on their exam grades. The BI-302 laboratory component complements the lecture and allows discovery and exploration of the topics.

## Academic Service Learning Project in Anatomy and Physiology: Peer Tutoring

Although the BI-302 lecture class was conducted as a traditional course, an extra credit service learning option to peer tutor provided students an opportunity to engage in constructivist learning. These students used the knowledge they had gained from their coursework to teach an authentic audience (those they tutored) for which learning this information matters. In addition, the peer tutors' process of reflecting in action and on action for this project was constructivist in nature, providing the mentor an opportunity to review and reflect upon prior and current knowledge with yet another experience. Students were offered an option of extra credit for peer tutoring. Tutoring assignments were designated after the first examination. Stronger students were assigned to tutor peers in their own classroom who did not perform well on the exam. Weaker students wishing to get extra credit also had tutoring assignments available to them. Keeping in mind that these weaker students had passed the earlier A&P courses, they were still competent enough to tutor A&P1 students, or pre-A&P (BI-110) students who needed help. In this way, all students participating in this service learning assignment were forced to review A&P to properly prepare for their tutoring. It is important to note that A&P2 students tutoring A&P1 students or pre-A&P students is a legitimate form of academic service learning since all topics taught in the earlier A&P courses are foundational for A&P2. As a result, even those students tutoring pre-A&P students were actively reinforcing knowledge that supports their current coursework. Students performing poorly in BI-302 and preliminary Anatomy & Physiology courses were provided tutoring by peers. Often peers better understand "sticking points" and can clarify conceptual difficulties. Peers also provide important mentoring that weaker students need in order to negotiate coursework. This type of assistance is invaluable for students struggling to pass these high-stakes courses. Receiving a decent grade in Anatomy & Physiology allows students to continue onto health careers such as nursing.

Once a BI-302 student is paired with another student needing tutoring, they arrange to meet outside of class time and decide on topics to review when they meet. For each hour of tutoring, the tutors receive one point of extra credit added to a single exam, up to a maximum of eight points/eight hours of tutoring. After each session, both the tutors and tutees were responsible for completing a reflective evaluation form that reviewed the details of each tutoring session as well as questions related to the success of the session (see Appendix). At the end of the semester, the service learning students in both the pre-A&P course and the A&P2 course were asked to complete an IRB-approved survey that allowed students to report on their perception of their own learning and their confidence level in a variety of skills important to success at school and in the workplace.

This study involved a range of science topics and project activities. The Academic Service Learning project alignment to course objectives are displayed in Table 8.3. The features for Physics, Biology and Anatomy and Physiology course projects are differentiated in Table 8.4 by mode of service delivery, course structure, assessment mode, course content focus, application of course content, level of interaction with service recipients, course peers and reflection protocols.

## ACADEMIC SERVICE LEARNING PROJECT SURVEY RESULTS

The first part of the survey asked students whether they agree or disagree with a list of statements as they pertain to the service learning component of the course in a 5-point Likert scale survey (i.e., strongly disagree, disagree, neutral, agree, strongly agree). We have combined the "strongly agree" and "agree" responses as a single positive response. In the second part of the survey students were asked to indicate if the service learning project helped them feel more confident about specific abilities as a result of their participation in service learning. In this portion of the surve, students responded using a 6-point Likert scale (not at all, a little bit, somewhat, quite a bit, a lot, not applicable). In the analysis "a lot" and "quite a bit" responses were combined

**TABLE 8.3 Academic Service Learning Alignment to Course Learning Objectives**

| | Course Learning Objectives | SL Activities That Support Student Learning Objectives |
|---|---|---|
| BI-110 Pre-Anatomy & Physiology | Students will develop a basic understanding of biological principles related to human body structure and function.<br>Students will analyze how disease impacts human anatomy and physiology.<br>Students will build a basic foundation of knowledge necessary for a career in health sciences. | To prepare for a poster presentation students must:<br>• research a disease<br>• compare normal human biology to anatomic and physiological changes to tissues and organs as a disease progresses<br>• assess current medical treatments<br>• prepare a handout for visitors listing prevention strategies |
| BI-302 Anatomy & Physiology 2 | Students will analyze the structure and functions of human body organ systems.<br>Students will evaluate the processes involved in homeostasis—the body's maintenance of a relatively stable internal environment.<br>Students will determine the factors that contribute to disease | As tutors, students must prepare on a weekly basis to teach peers about subject matter that includes:<br>• review of anatomical structure and function of various organ systems<br>• identifying homeostatic mechanisms that maintains the body's health.<br>• illustrations of how deviations from a state of homeostasis and/or disease processes cause disruption in normal cellular and organ functioning. |
| PH-101 Introduction to Physics | Students will develop a basic understanding of of concepts from various physics topics including:<br>• fforces and motion<br>• eenergy<br>• eelectricity and magnetism<br>• optics | To prepare for teaching a lesson on an assigned Physics topic students must:<br>• research the topic<br>• create activities to promote the discovery of Physics topics<br>• prepare materials to teach the topic |
| EDU-101 Contemporary Education | Students will study the ways in which changes in society have led to changes in educational goals, curriculum and instructional methods. | To prepare a Physics Friday lesson students must include<br>• related curriculum standards<br>• inquiry learning strategies<br>• activities to support diverse student needs<br>• assessment plan |

**TABLE 8.4 Differentiated Service Learning Implementation in Physics and Biology Courses**

| Service Learning Features | Physics Friday Inquiry Learning Activities | Health Fair Informational Presentation | Anatomy and Physiology Tutoring |
|---|---|---|---|
| Course Structure | Interdisciplinary<br>Physics/education<br>Learning community | Disciplinary<br>Biology | Disciplinary<br>Anatomy & Physiology |
| Assignment Mode | Physics: Extra credit option<br>Education: Required | Extra credit option | Extra credit option |
| Beneficiaries | | | |
| Providers | Preservice elementary education students | Prenursing students | Upper-level prenursing students |
| Recipients | K–6 students | High school students | Underachieving prenursing student peers |
| Focus | | | |
| Learning | Learning physics topics | Learning a disease | Reinforcing anatomy & physiology topics |
| Application | Self and others | Self and others | Self and others |
| Service | Provide community K–6 students with constructivist learning experiences in physics | Provide community high school students with information about disease | Assist struggling peers in learning required material in critical gateway courses |
| Level of Interaction with Recipients | High level of interaction in delivery of Inquiry learning activities in physics topics during the Physics Friday learning event | Interaction levels were dependent on participant interest and questions during the event | Multiple tutoring sessions with peer tutees throughout the semester |
| Level of Interaction with course peers to prepare service materials | Groups worked cooperatively with lab partners in physics course<br>Groups met to perfect inquiry method practice lessons in the first half of the semester, followed by creating physics topic lessons on assigned topics in Education course | Students worked independently on researching Biology course material and then prepared for their presentations in group settings | Students worked independently in preparing A&P material for tutees before each tutoring session. |
| Reflection | Repeated reflections on learning course content throughout the semester and pre and postservice learning | Reflections on application of course content to project throughout the project period; postdervice learning survey | Reflections on learning course content after each tutoring session; postservice learning survey |

as a single positive response. Survey items that related to the impact that the academic service learning projects had on learning, self-regulation, efficacy, and confidence in professional competencies were grouped by theme and listed in Table 8.5 to facilitate comparative analysis of students' percentage of positive responses for the different physics and biology projects.

## SURVEY DATA OF BIOLOGY STUDENTS

Students registering in the pre-A&P course were underprepared in the sciences. When asked if the service learning project helped them learn course material more effectively, and if they applied what they learned in class, their responses were about 20 points lower than the responses from students in the more advanced A&P2 course. This may result from the students' more minimalistic understanding of course material and how it relates to human health and disease. Alternatively, this difference in responses could be due to the project. In the A&P2 course, students peer tutored A&P, which directly aligns with course learning objectives. Preparing a health fair poster presentation only covered aspects of the course objectives.

When comparing responses related to self-regulated learning overall, students were fairly confident in their abilities. Both biology groups were highly confident that they could complete assignments on time, though the A&P2 students were more confident in their ability to follow directions and less confident in their ability to solve challenging problems. In critical thinking, responses were fairly similar between groups (less than 10% point spread between responses).

### Biology Student Reflections

Pre-A&P (BI-110) student reflection prompt: What did we learn in class that relates to your poster at the health fair? The students in the course wrote reflective comments about their experience creating posters for the health fair and how it impacted their learning. Many of the students' reflective statements were very specific about a precise topic in class that was strengthened as a result of preparing this health poster. Others responded with broader reflections such as the project "helped me learn more about what goes in our bodies and what causes changes and diseases we have." Another recognized that diseases may start in one place but can ultimately affect the entire body, writing,

> Complications from this disease can cause problem in other parts of the body because in class we learned that the body systems are all connected. If one

**TABLE 8.5 Impact of ASL Projects on Education and Nursing Students Science Learning, Self-Regulation, Academic Efficacy, and Professional Competencies**

| Academic Service-Learning Impact Areas | Survey Questions | % of Positive Responses | | |
|---|---|---|---|---|
| | | Education/Physics Interdisciplinary Learning Community ($N$ = 27) | Prenursing Program A&P, Within Class ($N$ = 17) | Prenursing Program AP-2 Between Class Levels ($N$ = 28) |
| Science Learning | The service learning project helped me learn material from my course more effectively | 81 | 65 | 82 |
| | I applied what I learned in class during the service learning experience | 89 | 65 | 86 |
| Self-Regulation | I am more confident in completing assignments on time | 89 | 94 | 92* (RR: 26/28)[a] |
| | I am more confident in following directions completely | 74 | 82 | 96* (RR: 26/28) |
| Efficacy | I am more confident solving challenging problems | 62 | 82 | 72* (RR: 24/28) |
| | I am more confident in asking questions in class | 63 | 76 | 87 (RR: 15/28) |
| | I am more confident analyzing and critically evaluating ideas, arguments and points of view | 89 | 88 | 81 (RR: 26/28 |
| | I am more confident comparing different approaches to solving a problem | 85 | 88 | 88 (RR: 26/28) |
| | I am more confident in expressing ideas, opinions, and facts verbally | 89 | 71 | 81 (RR: 26/28) |
| Professional Competencies | I have a good understanding of the community need my service learning project addressed | 70 | 53 | 71 |
| | The service learning project helped other people and or the community partner | 74 | 65 | 75 |
| | I am more confident to communicate with people from different backgrounds or cultures | 89 | 88 | 95 (RR: 15/28) |
| | I am more confident understanding the values of people different than me | 89 | 94 | 100 (RR: 15/28) |
| | I am more confident to work with others as members of a team | 93 | 88 | 93 (RR: 15/28) |

*Note:* RR = response rate.

> part of the body is affected, other parts would be affected too because the body systems work together.

This is a higher order of thinking about the disease process and one we typically hope students recognize in more advanced classes.

### A&P2 (BI-302) Reflection Prompt: Do You Think You Learned This Material Better Because You Had To Teach Someone Else?

Students provided positive qualitative statements indicating that they learned through the process of peer tutoring service learning. One student's comment demonstrates that students were filling holes in their A&P knowledge, writing, "I learned what I missed in Bi-301 and now [*sic*] understand that course material [*sic*]." The increased review of the course material piqued student interest as another student indicated, "Understanding the material makes me want to know more." Many of the collected reflective statements are eloquently summed up by this student, who wrote,

> What I really enjoyed about this method of studying is that having to prepare for the session made me memorize the material. Having to explain the information to someone else really solidifies the information you know, and gives you a different, more abstract way of thinking about it, especially when they ask questions and engage in the session.

## SURVEY DATA OF PHYSICS STUDENTS

The majority of students considered their overall experience in the service learning class as positive. The combined results of a survey of PH-101 students conducted at the end of the Fall 2013 and Spring 2015 semesters are listed on this table. Although the whole class prepared lesson plans, surveys were only collected from those students who participated in the service learning project and delivered a physics lesson to the after school academy students. The implementation details specific to those two semesters are summarized in Table 8.4. The Fall 2013 and Spring 2015 PH-101 sections were learning community classes that were linked to EDU-101 and combined in this table. All student participants in the Physics Friday ASL project were preservice teaching students. In questions related to course learning and connection, the majority of students found the service learning project relevant to the material learned in class with 81% of the students responding that the project did help them learn course material more effectively and 89% responding they applied what they learned in class during the service learning experience (Table 8.5).

On average 70%–80% of respondents reported having more confidence in their abilities in all of the categories. The smallest positive response rate overall was in the efficacy category measured by students rating related to gains in their confidence in "asking questions in class" (63%) and "solving challenging problems" (62%). Gains in confidence reported at 89% and upwards for this group were in completing assignments on time; analyzing and critically evaluating ideas, arguments, and points of view; expressing ideas, opinions, and facts verbally; communicating with people from different backgrounds; understanding the values of others; and working with others as members of a team.

Looking at survey responses from the service to community/partner perspective, the survey shows that 74% of the students agreed they believed the project helped the community member, while 70% understood the need that the ASL project addressed. In addition, there was a 30% increase in students' willingness to volunteer to do service in the community in the future.

Our community partner's feedback on the Physics Friday project was that it was a great success in that it greatly enriched the children's after school experience. They further noted that the science activities promoted a high level of engagement and enthusiasm in children who have exhibited disruptive behavior in the past. They are looking forward to our continued collaboration.

Students' ASL project reflections corroborate students' survey responses and provide greater insight into aspects of the project that contributed to the preservice elementary education students' responses.

## Physics Student Reflections

### *Reflection Prompt 1*

Describe central course concepts or theories you learned in your course that relate to your service learning project.

> I have benefited from the Physics Friday Science Lesson. It has helped me get a better understanding of my physics topic. I got to fully understand optics and how it works because I had to break it down for the students so that they could understand it. Also by going over the topic so many times it has fully stayed in my mind.

### *Reflection Prompt 2*

How did you observe/use/apply these course concepts or theories in your project?

Upon teaching, my understanding on Newton's third law I had to think of different situations in which I could enforce a child's understanding that for every action there is an opposite reaction. I am now able to understand that every action has a reaction, and furthermore am able to apply these new understandings in different situations. Such examples that correlate to such law are how a rocket, airplane function and much more.

### *Reflection Prompt 3*

Explain how your activities deepened your understanding of these course concepts or theories.

Upon having to create a lesson plan on my own on a given topic, and further having to teach students on it deepened my understanding on the given topic. Having to make the lesson simpler for the students of young age put me in a position where I had to think of the hands-on activities I was going to exercise, that would captivate their interest and at the same time would expand their knowledge. The activities I came up with to demonstrate the phenomenon I was given helped me further understand as well because this time instead of the professor giving me examples I came up with my own. I was able to get feedback from multiple perspectives from both the students and the professors. I had to use the knowledge of what I was taught in the phenomenon and my own personal knowledge to come up with a lesson plan.

### *Reflection Prompt 4*

Did your activities have an impact?

I believe that this Service-Learning project can promote my own learning process. Through teaching certain material my understanding on it expands, thus I fully need to understand a topic to further explain it to others. Since I need to create experiments to demonstrate a phenomenon I will have a long term understanding on the related topic.

This Physics Friday Science Lesson has also influenced the way I will teach science to my future students. I know that I have to make concepts as easy as I can and as fun as possible so that it will hold their interest and stay in their minds. I now know how to incorporate all kinds of different activities and how to engage the students in learning.

Now I feel better about science. I might not do so well on exams but I know in my head I get the information, even if it doesn't show on paper. With this confidence I feel better about teaching students science, because I know that if I can get it they will too. I didn't think I would be able to learn a physics concept so enough, much less to teach it, so my mind set has changed from the beginning of the semester to now. I went from being scared to know anything about physics to teaching it and really feeling satisfied that I got through kids heads and taught them something I didn't think I could learn myself.

> I definitely improved in science. I usually struggle with science but I did really well in Physics, the hardest of the science classes this semester. I always felt that teaching students science would be hard. It's hard for me to learn science in college so it has to be hard for a kid to learn science. But it's not impossible. With the right teaching method kids can be taught science.

## DISCUSSION AND CONCLUSION

The purpose of this study was to examine the impact that service learning projects had on the prenursing and teaching students' perceived ability to (a) learn material in required science courses more effectively, (b) apply course content during service learning experiences, and (c) have more confidence in self-regulated learning skills. Confidence in professional competencies was also analyzed. Some 72 professional career students in a culturally diverse urban community college in New York participated in this study. Of the 72 students, 27 were preservice elementary education students and 45 were prenursing students. All students were enrolled in science courses with dissimilar academic service learning projects in different course structures. This study also examined the relationships between the percentages of positive responses to survey questions about the service learning experience and the type of project in terms of course structure, activities, and level of interaction with service learning partners. In all, 17 of the prenursing student participants were enrolled in an entry-level Biology 110 (pre-A&P) course and 28 of the prenursing students were enrolled in an upper-level Anatomy and Physiology (A&P2) course. The 27 elementary education students were all enrolled in an Introduction to Physics (PH-101) course.

## PERCEIVED BENEFITS OF SCIENCE KNOWLEDGE

Although the academic service learning project activities were all aligned to science course objectives, analysis of the differences observed in students' survey responses may be sharpened through the lens of the social pedagogy design elements developed by Bass and Elmendorf (2012). One such dimension is the service learner's perception of the utility their science knowledge would have for their service learning partners. The prenursing program is highly rigorous, with early gateway courses and high-stakes assessments. The knowledge students must internalize is cumulative and must be learned to the point of automaticity through elaboration and multiple opportunities for application. For underprepared prenursing students, the effective teaching of scientific concepts plays an essential role in promoting the realization or disintegration of their career goals.

The service learning audience, or tutees in the A&P tutoring project, were poorly performing students enrolled in either Anatomy & Physiology 1 or 2 (A&P1 or A&P2), or in the pre-A&P course. In this project, prenursing students who were performing well in A&P2 were given the opportunity to revisit A&P fundamental information through their ASL task to assist their underachieving peers in A&P1 or pre-A&P. The project required students to use the knowledge they had gained from their coursework to teach an authentic audience and in essence review A&P1 material (which is foundational for A&P2) in preparation for their tutoring of A&P1 or pre-A&P students. The utility of preparing Anatomy & Physiology content material to assist their peers in passing this gateway course was clear and imminent.

In another ASL project, the prenursing students may have had a different perception about the utility of the science knowledge shared with the attendees of their health fair poster presentation. This ASL project was designed to afford students the opportunity to engage in an inquiry exercise to investigate the impact of particular diseases on normal body structure and function with the purpose of sharing this information with groups of high school students and the community at-large in a health fair. Formal instruction on body systems was delivered after students had experienced the ASL activities. Although the prospect of having to answer health fair participants' questions may have enhanced prenursing students' perceptions of their own understanding of the topic, the diverse nature of the large audience of health fair attendees may have caused the perceived utility of the information on a particular diseases less important a service to the visiting students. It is interesting to note that student reported confidence in "understanding of the community need addressed and people and/or community partner helped" was the lowest reported positive response, at 53%, for this project, while preservice teaching Physics students and Anatomy & Physiology students involved in teaching and tutoring ASL projects generated a 70–71% positive response.

In the PH-101 project, Physics Friday, preservice teachers prepared inquiry learning activities to teach elementary-aged students fundamental concepts from their introductory Physics course. The utility of the content material in this case was threefold. For the preservice teaching students enrolled in the Physics and Education learning community, learning outcomes in science and pedagogy were aligned to the ASL project. Thus, both science topic concepts and best practice strategies of teaching and inquiry learning were applied in the ASL lesson preparation and delivery. The benefits were reciprocal and served to reinforce the knowledge of the service learner and enrich the science learning experiences of the recipients of the service. In comparing the impact of the three ASL projects on student learning, the A& P tutoring project and Physics Friday teaching project percentages of positive student responses were on a par with each other and

20% higher in positive responses than the health fair presentation project for understanding the service learning partners' needs and learning and applying science course material more effectively. The direct application of science content in promoting immediate academic success was present in PH-101 and the upper sequence A&P1 and 2 courses.

## PERCEIVED IMPORTANCE OF COMMUNICATING UNDERSTANDING OF SCIENCE KNOWLEDGE

Another difference identified in the academic service learning projects in this study was the prenursing and preteaching student's perception of the importance of their ability to effectively communicate their understanding of their respective science concepts to their respective audiences. The audience for the A&P tutoring project was in the form of a single struggling peer. This dynamic caused students to deepen their understanding with their tutees in mind. One tutor reflected,

> Studying with a partner where you are responsible for tutoring them on the material is somewhat nerve racking. Even with being confident in my competency of the topic, I fear I may be confusing something or making a topic not clear.

The motivation for attaining a deeper level of understanding of Physics course concepts to then effectively teach to their service learning students was further elaborated in this Education student's reflection on the project, "It has helped me get a better understanding of my Physics topic. I got to fully understand optics and how it works because I had to break it down for the students so that they could understand it." The prospect of their science knowledge being tested in an open-ended context sparked the realization that a surface-level knowledge of their topics might be inadequate for encouraging the questions of their students as they move through the stages of the inquiry lesson plan. Whereas in the BI-110 service-learning activity, preservice nursing program students conducted research on diseases which may or may not have direct impact on their intended service learning partners' needs.

In summary, the data suggest the academic service learning project assignment structures which provided students with opportunities to learn course science material with a focus on their service learning audience and awareness of the value of their representation of that knowledge to the specific needs of their service learning partners resulted in higher percentages of positive responses in science learning survey questions and an understanding of the need the project addressed. The projects whose service learning audience was more abstract in nature generated lower positive student responses. The value of representing the knowledge for such a diverse audience was therefore ill-defined. Positive self-regulated learning confidence responses, upwards of 90%, were highest among prenursing student tutors for time management and following directions, with prenursing student presenters reporting high levels of time management skills as well. The students reporting the highest percentages of positive responses in the area of confidence in communicating with and understanding the values of culturally diverse people served their struggling peers through repeated individualized tutoring sessions on high-stakes gateway course content in Anatomy & Physiology. The highest positive response percentages reported by elementary education students were in Physics learning application (89%) and collaborating with others as members of a team (93%). These results corroborate the research findings of a study conducted by the Science Education Outreach Program (Gutstein, Smith & Manahan, 2006), where undergraduates applied their knowledge of contemporary educational theories and methods in the context of science education in community schools. A review of the program indicated that service learning students had applied the information they learned and expressed that the project had positive impacts on their education and career choices and workplace skills.

## IMPLICATIONS

This research sought to explore the ideal that all students can achieve understanding of science if given the opportunity to learn through effective research-based practices. Preprofessional prenursing and elementary education students in science courses required by their programs engaged in learning science material through a catalytic high-impact practice—academic service learning. This practice supports the current science reform movement's call for a substantive change in preprofessional development and in how science is taught. This approach included experiences that engaged teaching and nursing students in constructivist learning activities that promoted their knowledge of science course content and applied that knowledge in an activity with an authentic audience. The ASL project exemplified the benefits of improved instruction and informed curricular and pedagogical approaches called for by leading educator Diane Ravich (2010).

The findings in this study suggest that the affective domain component in the ASL activities may have contributed to the upwards of 20-point increase in positive responses to the question of whether the ASL project positively impacted students' learning and application of Science course content. Astin, Vogelgesang, Ikeda, and Yee (2000) conducted a longitudinal study on how service learning affects students and similarly found that the most important factor related to a positive service learning experience was whether the service promoted an understanding of the academic course material. The data also showed that students benefit from the opportunity to link the practical service experience to the intellectual concepts in the classroom. Future iterations of the ASL projects can be designed with the intention of connecting affective learning objectives to service learning experiences identified in Hall's pioneering work (2006). Affective Domain outcomes may be targeted through service learning pedagogy by design to promote (a) receiving/attending to someone or something; (b) responding—taking action, giving feedback, obtaining satisfaction; (c) valuing—commitment to something because of its perceived worth or value; (d) organization—adding new values or beliefs and beginning to organizing life around those values; and (e) characterization—integration of values into daily lifestyle or philosophy (Krathwhol, Bloom, & Masia, 1964). The social pedagogy design elements outlined in the work of Bass and Elmendorf (2012) provide service learning project planners with powerful design strategies to create opportunities for prenursing and preservice elementary education students to learn science by connecting their affective and cognitive domains through activities which require them to communicate their knowledge of course content to authentic audiences for whom the information matters.

## APPENDIX
## Anatomy and Physiology ASL Tutor Report

Name ____________________

Name of Study Partner ____________________

*Report Sheet for BI-302 PNET: Extra Credit Assignment:*

Date of meeting: ____________________

Meeting was from __________ until __________

Your Signature ________________ Signature of Partner ________________

1. What topics did you prepare for this session? What topics did your partner prepare?
2. How much time did you spend preparing? Did you feel confident you understood the information you need to teach?
3. Did your study partner come prepared for the session? Explain.
4. Do you feel you equally shared the responsibilities for this study session? Please explain.
5. What went well with the session? What can be done to improve the study sessions?
6. Do you think you learned this material better because you had to teach someone else?

## REFERENCES

Astin, A. W. (1993). *What matters in college: Four critical years revisited.* San Francisco, CA: Jossey-Bass

Astin, A. W., & Sax, L. J. (1998). How undergraduates are affected by service participation. *The Journal of College Student Development, 39*(3), 251–263.

Astin, A. W., Vogelgesang, L. J., Ikeda, E. K., & Yee, J. A. (2000). *How service-learning affects students.* Los Angeles: Higher Education Research Institute, University of California.

Bass, R., & Elmendorf, H. (2012). Social pedagogies: A Teagle foundation white paper. Lord's constructivist model put to a test. *American Biology Teacher, 65*(7), 491–502.

Beichner, R., Saul, J., Abbott, D., Morse, J., Deardorff, D., Allain, R., . . . Risley, J. (2006). Student-centered activities for large enrollment undergraduate programs (SCALE-UP) project. In E. F. Redish & P. J. Cooney (Eds.), *PER-based reform in university physics.* College Park, MD: American Association of Physics Teachers.

Burrowes P. A. (2003). A student-centered approach to teach general biology that really works: Lord's constructivist model put to a test. *American Biology Teacher, 65,* 491–502.

Chrispeels, H. E., Klosterman, M. L., Martin, J. B, Lundy, S. R., Watkins, J. M., Gibson, C. L., & Muday, G. K. (2014). Undergraduates achieve learning gains in plant genetics through peer teaching of secondary students. *Life Sciences Education, 13,* 641–652.

Cooperstein, S. E., & Kocevar-Weidinger, E. (2004). Beyond active learning: A constructivist approach to learning. *Reference Services Review, 32*(2), 141–148.

Coulter, T., & Vandal, B. (2007). *Community colleges and teacher preparation: Roles, issues and opportunities.* Denver, CO: Education Commission of the States.

Dewey, J. (1933). How we think: A restatement of the relation of reflective thinking to the educative process. Boston, MA: D.C. Health.

Duschi, R. A., Schweingruber, H. A., & Shouse, A. W. (2007). *Taking science to school: Learning and teaching science in grades K–8.* Washington, DC: National Academies Press.

Ellerton, S., DiMeo, C., Kemmerer, A., Pantaleo, J. , Bandziukas, M., Bradley, M., & Fichera, V. (2014). Service-learning as a pedagogical tool for career development and vocational training. In A. Traver & Z. P. Katz (Eds.), *Service-learning at the American community college: Theoretical and empirical perspectives* (pp. 211–224). New York, NY: Palgrave Macmillan.

Engstrom, C., & Tinto, V. (2008, January/February). Access without support is not opportunity. *Change: The Magazine of Higher Learning, 40*(1), 46–51. Retrieved from http://www.changemag.org/Archives/Back%20Issues/January-February%202008/abstract-access-without-support.html

Evelyn, J. (2002, March 8). States give community colleges a role in educating teachers. *The Chronicle of Higher Education, 48*(26) 22–23.

Eyler, J., & Giles D. E. (1999). *Where's the learning in service-learning? Jossey-Bass higher and adult education series.* San Francisco, CA: Jossey-Bass.

Fathe, L.A., Kasabian, J. (2008). *The state of affairs: Impact and implications of STEM teacher education at two-year colleges.* Torrance, CA: El Camino Community College District.

Freeman, S., O'Connor, E., Parks, J. W., Cunningham, M., Hurley, D., Haak, D., & Wenderoth, M. P. (2007). Prescribed active learning increases performance in introductory biology. *CBE-Life Sciences Education, 6*(2), 132–139.

Fulcher, R. (2002). Community colleges are primary educators of RNS, meet area needs. *Legislative Network for Nurses, 19*(24).

Furco, A. (1996). Service-learning: A balanced approach to experiential education. *Expanding boundaries: Service and learning* (pp. 2–6). Washington, DC: Corporation for National Service. Retrieved January 2, 2016, from https://www.fresnostate.edu/academics/cesl/documents/facultysl/facultysl-ABalancedApproach.pdf

Good, T. L., & Brophy, J. E. (1994). *Looking in classrooms.* New York, NY: HarperCollins.

Gutstein, J., Smith, M., & Manahan, D. (2006). A service-learning model for science education outreach. *Journal of College Science Teaching, 36*(1), 22–26.

Hake, R. R. (1998). Interactive-engagement versus traditional methods: A six-thousand-student survey of mechanics test data for introductory physics courses. *American Journal of Physics, 66*(1), 64–74.

Hall, M. (2006). Service-learning and intentionality: Creating and assessing cognitive affective learning connections. *Journal of Higher Education Outreach & Engagement, 11*(2), 33–43.

Hamlernick, J. (2013). *Strengthening STEM education through service-learning (2013). Highlights from 2010 Learn and Serve America STEM grants.* Corporation for National Service. Washington, DC: Retrieved from http://mncampuscompact.org/wp-content/uploads/sites/30/2016/04/strengthening-STEM-through-SL-pdf

Hark, A. T. (2008). Crossing over: An undergraduate service learning project that connects to biotechnology education in secondary schools. *Biochemistry and Molecular Biology Education, 36*(2), 159–65.

Harris, D. E., Hannum, L., & Gupta, S. (2004). Contributing factors to student success in anatomy & physiology: Lower outside workload & better preparation. *The American Biology Teacher, 66*(3), 168–175.

Honnet, E. P., & Poulsen, S. J. (1989, October). Principles of good practice for combining service and learning. *University of Nebraska, Omaha.* Retrieved from http://digitalcommons.unomaha.edu/cgi/viewcontent.cgi?article=1001&context=slceguides

Hopp, R. M. P. (2009). A success story: Chemistry before anatomy and physiology I. *HAPS Educator,* 56–59.

Institute of Medicine. (2011). The future of nursing: Leading change, advancing health. *National Academies Press.* Retrieved from http://nationalacademies.org/hmd/reports/2010/the-future-of-nursing-leading-change-advancing-health.aspx

Iverson, H. L., Briggs, D., Ruiz-Primo, M., Talbot, R., & Shepard, L. A. (2009). Physics course: Innovations and their impact on student learning. *AIP Conference Proceedings 1179,* 169–172.

Knight, R. D. (2002). *Five easy lessons: Strategies for successful physics teaching.* San Francisco, CA: Addison Wesley.

Kostovich, C. T., Poradzisz, M., Wood, K., & O'Brien, K. L. (2007). Learning style preference and student aptitude for concept maps. *Journal of Nursing Education, 46,* 225–231.

Krathwhol, D. R., Bloom, B. S., & Masia, B. B. (1964). *Taxonomy of educational objectives: Handbook II: Affective domain.* New York, NY: McKay.

Kuh, G. (2008). *High-impact educational practices: What they are, who has access to them, and why they matter.* Washington, DC: Association of American Colleges and Universities.

Lewis, E., Dema, O., & Harshbarger, D. (2014, April 1). Preparation for practice: Elementary preservice teachers learning and using scientific classroom discourse community instructional strategies. *Wiley Online Library.* Retrieved from http://onlinelibrary.wiley.com/doi/10.1111/ssm.12067/abstract

Lumpe, A. T., & Staver, J. R. (1995). Peer collaboration and concept development: Learning about photosynthesis. *Journal of Research in Science Teaching, 32,*71–98.

Lutz, S. T., & Huitt, W. G. (2004). Connecting cognitive development and constructivism: Implications from theory for instruction and assessment. *Constructivism in the Human Sciences, 9*(1), 67–90.

Martin, D. J., Jean-Sigur, R., & Schmidt, E. S. (2005). Process-oriented inquiry: A constructivist approach to early childhood science education: Teaching teachers to do science. *Journal of Elementary Science Education, 17*(1), 13–26.

Michael, J. (2006). Where's the evidence that active learning works? *Advances in Physiology Education, 30*(4), 159–167.

Mullis, I., V. & Jenkins, L. B. (September 1988). *The science report card: Elements of risk and recovery.* Princeton, NJ: Educational Testing Service.

National Research Council. (1996). *National science education standards.* (1996). Washington DC: National Academy Press. Retrieved from https://www.nap.edu/read/4962

Niemi, H. (2009). Why from teaching to learning? *European Educational Research Journal, 8*(1), 1–17.

Nyatanga, I. (2005). Nursing and the philosophy of science. *Nurse Education Today, 25*(8), 670–674.

Othmer, K., & Sealfon, C. (2010). Integrating an elementary school service-learning component into a college physics course for non-majors. *Science Education and Civic Engagement, 2*(2), 29–34.

Plourde, L. A. (2002). The influence of student teaching on preservice elementary teachers' science self-efficacy and outcome expectancy beliefs. *Journal of Instructional Psychology, 29*(4), 245–253.

Prentice, M., & Robinson, G. (2010). Improving student learning outcomes with service-learning. *Digital Commons @ UNO* Retrieved from http://digitalcommons.unomaha.edu/slcehighered/148

President's Council of Advisors on Science & Technology. (2010). *Prepare and inspire: K–12 Education in science, technology, engineering, and math (STEM) for America's future: Executive report.* Washington, DC: Author.

Queensborough Community College (QCC). (2014). *2014–2015 fact book.* Retrieved from http://www.qcc.cuny.edu/oira/docs/factbook-2014-15/FactBook2014-2015.pdf

Radford, A. W., Berkner, L., Wheeless, S. C., & Shepherd, B. (2010). *Persistence and attainment of 2003–04 beginning postsecondary students: After six years* (NCES 2011151). Washington, DC: U.S. Department of Education, National Center for Education Statistics.

Ravitch, D. (2010). In need of a renaissance: Real reform will renew, not abandon our neighborhood schools. *American Educator, 34*(2), 10–22.

Richardson, J. T. E. (1994) Mature students in higher education: a literature survey on approaches to studying. *Studies in Higher Education, 19*(3), 309–325.

Roberts, M. W. (2000). Birth of the catalytic concept (1800–1900). *Catalysis Letters,* 67(1), 1–4.

Rodgers, C. R. (2002). Seeing student learning: Teacher change and the role of reflection. Voices inside schools. *Harvard Educational Review, 72*(2) 230–253.

Schon, D. A. (1983). *The reflective practitioner: How professionals think in action.* New York, NY: Basic Books.

Schon, D. A. (1987). *Educating the reflective practitioner.* San Francisco, CA: Jossey-Bass.

Shkodriani, G. (2004). *Seamless pipeline from two-year to four-year institutions for teacher training.* PT3 Policy Brief. Denver, CO: Education Commission of the States.

Shrigley, R. L. (1983). The attitude concept and science teaching. *Science Education, 67,* 425–442.

Shulman, L. (2004). *The wisdom of practice: Essays on teaching, learning, and learning to teach.* San Francisco, CA: Jossey-Bass.

Sigmon, R. L. (1997). Linking service-learning in liberal arts education. Retrieved December 3, 2015, from http://eric.ed.gov/?id=ED446685

Smith, A. (March 26, 2015). Nearly half of four-year college students attended two-year college. *Inside Higher Education.* Retrieved from https://www.insidehighered.com/users/ashley-smith?page=25

White, J. (1999). Wellness Wednesdays: Health promotion and service learning on campus. *Journal of Nursing Education, 38*(2) 69–71.

Wilke, R. R., & Straits, W. J. (2001). The effects of discovery learning in a lower-division biology course. *Advanced Physiology Education, 25,* 62–69.

CHAPTER 9

# LEARNING AND EMOTION

## The Power of Service Learning in STEM Fields

**Maureen P. Hall and Aminda J. O'Hare**
***University of Massachusetts Dartmouth***

### ABSTRACT

This investigation explores service learning as an experiential conduit for learning in context where real-world problems and emotions can be dynamically combined and where emotions and learning can be knitted together successfully. Service learning, as a kind of experiential education, provides a needed foil to the current overemphasis on standardized testing in education. This hyperfocus on testing ignores the development of the whole person in authentic learning contexts and overlooks the neuroscience of teaching and learning. A case study of a service learning intervention with teacher education students is reported in this chapter and grounds this work in practice, showcasing the possibilities for enlarging pedagogical learning opportunities in STEM fields. Research from cognitive psychology and cognitive neuroscience is cited to help build the case for emotions in learning. Through engagement and reflection on engagement through service learning contexts

*Science and Service Learning*, pages 221–239

in STEM fields, optimal levels of engagement can be achieved and students can understand more deeply the broader impacts of the STEM disciplines.

Current education climates privileging teaching toward standardized assessments of content comprehension do not provide students with a framework through which to apply their knowledge to real-world situations. While deepening and broadening one's understanding of their chosen field is an essential part of becoming an educated person, that knowledge is meaningless unless its applications in the field are also understood. An educated person is able to communicate ideas with nonexperts and understand the impact of decisions made on larger communities. Incorporating service learning experiences into higher education coursework provides a conduit for students to understand what it means to be an educated professional outside of the classroom, helps them to understand real-world problems, and enriches their understanding of the broader impacts of their disciplines. This is particularly true in science, technology, engineering, and mathematics (STEM) curricula where students may have a hard time connecting to the course material. This chapter discusses how service learning experiences engage affective learning mechanisms. These effects will be demonstrated through a service learning project conducted with preservice educators. The application of these effects to broader STEM disciplines will also be discussed.

## Service Learning: Learning in Context and Through Reflection

As an approach connected to the learning goals of a college or university course, service learning provides a dynamic conduit for learning in context; one where real-world problems can be examined authentically and built into the course curriculum. Service learning, by definition, responds to a community need, but it is not simply community service. Service learning provides opportunities for students to give service and then to integrate the experience back into the curriculum of a college or university course (Hall, 2009). What is the mechanism or process of learning in service learning? Eyler (2001) argued that "reflection is the hyphen in service-learning" (p. 35). Reflective practice, including reflective thought, reflective writing, and dialogic inquiry with others, becomes the artifacts of the actual learning taking place in the individual (Hall, 2012). This is where Eyler's "hyphen" is evidenced; the learning is activated through reflection.

The act of intentional reflection and reflective practice can help to uncover the connections between one's service and to one's own learning possibilities. William Cronon (1998), in an essay for *The American Scholar*, underlined the importance of both noticing and understanding linkages as a critical component of the process of becoming educated: "More than anything else, being an educated person means being able to see connections that allow one to make sense of the world and act within it in creative ways" (p. 78). Cronon also suggested that education should aspire "to nurture the growth of human talent in the service of human freedom" (p. 83).

Service learning, as a dynamic example of experiential education, takes place in social contexts and holds promise for creating learning pathways to and from the classroom and the real world. As McKay-Jackson (2014) pointed out, "Traditional schooling in reading, writing, and arithmetic alone will not fully prepare students to lead with integrity, govern fairly, analyze problems, and work collectively with people different from themselves" (p. 293). Part of the promise of service learning involves connecting affective and cognitive learning, thereby connecting learning and emotion in a way that can help to develop the whole person. It is important to consider affective learning as involving more than just emotions. The human brain does not separate emotions from cognitions. Without any attention given to students' interest, motivation, appreciation, and attitudes, real and enduring learning is incomplete (Chickering, 2006; Owen-Smith, 2004). Neuroscientists Immordino-Yang and Faeth (2009) pointed out that no learning can any longer be considered separate or disrupted by emotion. Instead, they argued, "like other forms of learning and interacting, building academic knowledge involves integrating emotion and cognition in social context" (p. 2).

What we must keep coming back to is the question, "What does it mean to be educated?" One important answer to this question is that education helps to prepare students to become active and reflective members of a democratic society. Service learning stands as one important route to learning as it integrates the individual into the community then brings the community back to the classroom for synthesis and extended or deepened learning (Hall 2007).

## CHALLENGES IN EDUCATION AND A CALL FOR CHANGE IN EDUCATIONAL APPROACHES

Within the first 3 to 5 years of entering the teaching profession a 30% attrition rate occurs (Darling-Hammond, 1998). Anderson and Olson's (2007) work in urban teacher retention highlighted the need to reconceptualize teacher retention. Their findings revealed that urban teachers will remain in urban education if they can adopt multiple education roles inside and outside of the classroom and receive professional support during the whole of their careers instead of just at the beginning (Anderson & Olson, 2007).

Higher education professors and K–12 teachers today require a multitude of skills to produce critical thinkers and problem solvers (Meiklejohn et al., 2010). Despite extensive efforts by teachers, our current system of education, with its hyperfocus on high-stakes standardized testing, has not led to success for all. Without the integration of mind, body, and spirit into the spaces of teaching and learning, the classroom becomes a place promoting dehumanization and doing a disservice to the human beings who inhabit those spaces. Schools may teach students, but they do not help students find wisdom or insight.

Increasingly in education, what seems to matter most are students' scores on standardized tests, although success on these tests is often the product of rote memorization rather than of critical thinking and problem-solving. This lends to fragmentation and disconnection in the spaces of teaching and learning. This hyperfocus ignores the development of the whole person, but also the development of creativity and the power of imagination. We identify at least two outcomes of our current education environment focused on standardized testing that are creating isolated, deindividualized students: training memorization skills over critical thinking skills, and deemphasizing students' development of passion and joy for learning.

Darling-Hammond (2007) discussed how standardized testing affects K–12 pedagogical practices in the classroom, pointing out that "Teachers increasingly report that the curriculum is distorted by tests and that they feel pressured to 'teach to the test' in ways that contradict their ideas of sound instructional practice" (p. 326). Educator Bobbie Solley (2007) elaborated on the ways in which standardized testing affects instructional approaches and skills:

> Important skills that schools once taught, such as critical thinking, discussions, and problem-solving, are being replaced by low-level, fill-in-the-blank worksheets and drills... It is past time for teachers, schools, administrators, parents, and the public to stand up and let their voices be heard. It is past time for all involved to put a stop to the *misuse* of tests in all grades. (p. 31, emphasis in original)

Standardized testing as a kind of external accountability actually functions as a way to deskill teachers and to deprofessionalize teachers' roles (Shepard, 2000; Solley, 2007).

High-stakes accountability teaches students that effort in school should be in response to externally administered rewards and punishment rather than to the excitement of ideas. This view degrades learning itself and dehumanizes human potential. Research in economics and psychology have long shown external sources of motivation narrow attention, stunt creativity, and impact behavior only in the short-term, while internal sources of motivation broaden attention, increase creativity and problem-solving, and impact behavior in the long-term (for a review, see Bénabou & Tirole, 2003). Unlike the hyperfocus on standardized testing, teaching that encourages development of the whole human being—something that holistic education attends to—seems much more just, humane, and well-suited to the education of human beings.

In every educational venue and level, there is a call for renewed perspectives and practices on knowing and living in contemporary education, echoing Arthur Zajonc (2006): "I am calling for resituating it [education] within a greater vision of what knowing and living are really all about. That re-imagination of knowing will have deep consequences for education" (p. 1743). In this passage, Zajonc emphasized the importance of renewing and a "reimagination of knowing" as a process of ongoing transformation and as a way of gaining both self-knowledge and knowledge of others. Likewise, new interdisciplinary, multidisciplinary, and transdisciplinary approaches in teaching and research can provide routes to transforming education (Klein, 2005). Inter- or multidisciplinary approaches privileging both cognitive and affective gains will help promote justice through curricular approaches by taking into account the mind, body, and spirit of both students and teachers. As we consider what it means to educate and to be educated, we must also consider ways to include all voices.

## THEORETICAL FRAMEWORKS UNDERPINNING THE WORK OF SERVICE LEARNING

The evolving work of this project draws upon at least two main theoretical frameworks: constructivism—more specifically, an outgrowth of the constructive developmental theory of teaching and learning, and the framework of transformational change (Eckel & Kezar, 2003). As characterized in Marcia Baxter Magolda's (2000) work, the constructivist developmental theory incorporates two major concepts: "(1) students construct knowledge by organizing and making meaning of their experiences, and (2) this construction takes place in the context of their evolving assumptions about knowledge itself and students' role in creating it" (p. 6). As an outgrowth and practical application of this theoretical underpinning, both cognitive and affective dimensions of learning are privileged in the best teaching and learning spaces that utilize service learning. For authentic learning, there must be some kind of hands-on or practical application of what a student has been learning in a conceptual manner (Berman, 2006) as provided by service learning. There is a central belief in the importance of teaching to and from the whole person (Palmer, 1998), along with an understanding that professors cannot make students learn. Through practical applications of learning and reflecting on that application, students are provided with rich learning possibilities.

While the first theoretical framework focuses on the relationships between and among teachers and learners, the second theoretical framework involves change in the larger institution. The framework of transformational change (Eckel & Kezar, 2003) presents a proactive rather than a reactive process to affect transformation. Service learning and reflection on that learning holds promise to operationalize parts of both the constructivist developmental theory and the theory of transformational change.

We would be remiss not to also draw upon the ideas of John Dewey (1899), one of the 20th century's foremost educational thinkers. Dewey emphasized that reflection creates the necessary connections between experience and theory. Self-reflection, however, has been emphasized through the ages. The ancient Greeks urge us, "know thyself." Socrates warned that an "unexamined" life lacks meaning; experience without reflection on experience is meaningless. The act of examining our lives as a text is a critical process. Learning how to be a critical thinker is certainly one part of being educated. Long before the Greeks, ancient Indian wisdom called for the same self-knowledge. Ancient Indian texts, including the Vedas and the Upanishads, contain kernels of wisdom that can be adapted and applied to the best education for teachers in the 21st century and beyond. Prominent Indian philosopher Aurobindo articulated one of the main principles of education, the "mind has to be consulted in its own growth.... It is he himself who must be induced to expand in accordance with his own nature" (1910). Aurobindo privileges self-reflection and reflective practice as an important conduit for learning to happen. To give additional meaning to our pedagogy, we also drew on the conceptual frameworks of Lev Vygotsky (1978) and Jean Piaget (1954), who represent the early development of constructivist theory, one which places the student at the center of the learning experience and gives the teacher the role of providing an environment focused on how the students can best learn and make meaning (Davis & Sumara, 2002).

Part of Dr. Maureen Hall's early work with service learning grew out of her membership in the Carnegie Leadership Program over 2006–2009. The focus of this work was studying and researching the connections between cognitive and affective learning using the Scholarship of Teaching and Learning (SoTL) as a conduit or method. The Carnegie Leadership Program was program supported by the Carnegie Academy for the Scholarship of Teaching and Learning (CASTL). Hall's work in CASTL centered on the cognitive and affective effects of service learning on preservice teachers. She was the initiator and contact leader at University of Massachusetts Dartmouth for this 3-year Carnegie Leadership Program Group focusing on Cognitive-Affective Learning (CAL) in student learning. More recently, as a member of the STEM Education and Teacher Development Department at the University of Massachusetts Dartmouth, she has started to collaborate on research projects with cognitive neuroscientist Dr. Aminda O'Hare in investigating the interaction of learning and emotion and how this interaction plays a role in improving service learning pedagogy. In this chapter we document Maureen's early work with service learning and her current work with Aminda as we explore the role of reflection inherent in service learning. We further describe how the synergy between learning and emotion can deepen students' understanding of themselves, others, and the applications of their course content. This chapter explores possibilities for how service learning and the reflection connected with service learning can impact people who are working toward STEM careers.

## CASE STUDY OF HALL'S COGNITIVE AFFECTIVE LEARNING WITH SERVICE LEARNING

The participants in Hall's service learning study in 2009 were enrolled in the introductory course leading to teacher licensure. Interactive in nature, this course was designed to introduce potential teachers to the real world of teaching through an in-depth, candid analysis of the teaching profession, including an exploration of the challenges and rewards of teaching. The history, philosophy, sociology, and politics of American education were studied as were the current educational issues, trends, and reform movements. It was the first course in the sequence required for teacher licensure. The students in this course arrived with many notions of what it means to be a teacher. Most of them had had an array of wonderful teaching and learning experiences, were self-motivated, and came to class ready to learn. Many of these preservice teachers had never encountered students who lacked motivation and positive role models or did not enjoy learning. To bring about deep and enduring learning about the teaching profession, the course utilized a service learning component. Students learned theory at the university then got real-world K–12 experiences at an urban alternative school to see if and how these theoretical pieces translate into the world of practice (Ball & Cohen, 1999). The strength of service learning is making the theoretical visible through a practical experience.

Data were generated through the service learning component of this introductory course. The artifacts for this particular study included the data collected from preservice teacher questionnaires (at two points in each semester) and from the reflective component of end-of-semester case studies on urban alternative school students. The questions and criteria for each of these teaching artifacts are provided in the appendices of this chapter. Making this work public added to the literature on preservice teacher education and invited other teacher scholars to build upon it. Lee Shulman, former President of the Carnegie Foundation for the Advancement of Teaching, points out that unlike fields such as architecture and law, where new work builds on a body of existing work, teaching is devoid of a history of practice (Ornstein, Behar-Horenstein, & Pajak, 2003).

## METHOD

Qualitative data were collected to investigate how a service learning component might inform and influence students' decisions to become teaching professionals. For this course, in line with both the course and the state licensing agency's guidelines, students were required to complete 15 hours of tutoring in order to fulfill course and state requirements for prepracticum hours. These hours were simultaneously considered service learning since university students worked to meet the needs of the K–12 community then reintegrated these experiences back into the university course, making connections and disconnections to the research base and theoretical framework of the teaching profession. Students' experiences were documented in reflective diaries and student interview questions. By exploring their service learning experiences, students articulated their notions of the teaching profession as a whole. They shared values, interests, and motivations, thereby activating the affective domain of learning. Data were collected over the course of 2 years. The participants in this study were assigned to work as tutors to individual seventh and eighth graders at an alternative urban secondary school. Many students in this setting had been previously expelled from mainstream public schools for various reasons, including behavior and discipline problems. Data also were collected to capture preservice teachers' experiences in terms of affective development in student learning using two methods; first, through the implementation of a questionnaire composed mainly of affective-based questions about their prepracticum experiences, which involved tutoring and teacher observations at this urban school site. The questions were designed to allow data regarding students' motivation and interest in becoming a teacher. See the questionnaire (Appendix A) administered to preservice teachers twice over the course of each semester.

The second method of data collection was to use the reflective piece of the assigned case study, an integral part of the coursework for all participants in the study. The portion of the case study involved a reflective piece on Parker Palmer's (1998) *The Courage to Teach* in relation to their service learning experiences at the urban alternative school. Preservice teachers were asked to identify and write about an important theme from Palmer's text articulating that theme in connection with the student they tutored at the urban alternative school. These data represented another way to document the learning in service learning. Palmer's text was one of three required texts used for this course. It is important to note that all teachers at this alternative high school were also reading Palmer's text, which raised awareness in interest and motivation for learning.

## DATA ANALYSIS

An inductive process was utilized for the first set of data, where emerging themes were generated directly from student voices. The data were systematically gathered through the research process then analyzed. This analysis refers to sense-making of the collected data and an articulation of the themes emerging from the data. In writing about qualitative data analysis, Miles and Huberman (1994) stated that the focus is on data in the form of words. In this case, words emanate from the interviews conducted, and the processing of these words and ideas represent the form of analysis used.

The first dataset grew out of the student interview questions (see Appendix A), which were administered to students twice in each of the two semesters. Data were collected in the 4th week and then again in the 12th week of each semester. By the 4th week of the semester, each participant had begun the tutoring of his or her student at the alternative school. By the 12th week, each participant had had the opportunity to meet with the student being she tutored several times. Each participant spent 15 hours with the student, and by the 12th week, more than 10 hours of tutoring had occurred. Numbers were randomly assigned to each participant, ensuring anonymity for the participants and allowing a view of each student's progression from the first data collection to the second.

To construct the second dataset (see Appendix B), participants chose a theme of interest from Chapters 1 and 2 of the Palmer text (1998). Each participant wrote about the meaning of that theme in the context of their tutoring experiences at the alternative school. This part of the study had a more deductive approach because participants had to choose an existing theme from Palmer's work and make meaning of that theme in their teaching lives. This part of the study aimed at deepening the understanding of identified themes of importance to preservice teachers. It was an attempt to get an inside view of participants' ideas of the relevance of Palmer's themes to their current and future teaching lives. This model of investigation provided a way to encourage a more in-depth understanding involved in entering the teaching profession.

## RESULTS

In the first section of the findings, the emerging themes are organized by student interview question. These interview questions were anonymous and given to two different groups of students from this course. Under each of the questions below, emerging themes are described, with specific examples representative of students' responses. In the second section, data from the reflective component of the assigned case study are reported. Similarly to section one, emerging themes are listed with accompanying quotes culled from students' case studies.

## Part One: Responses to Preservice Teacher Interview Questions

Question 1 considered students' perceptions of the teaching profession as a whole. Three main themes emerged from this student data: (a) teaching as a challenging profession, (b) teaching as an undervalued profession, and a (c) lack of respect. Question 2 elicited students' positive experiences, informing their interest in becoming a teacher. One main theme, making a difference, emerged in the data. Students witnessed the power of teachers' influence on the course of young lives and this engaged and fostered interest in the profession. Question 3 focused on elements in students' service learning experiences that detracted from their interest in becoming a teacher. Two main themes emerged in students' responses: lack of respect, and apathy, disruption, and lack of motivation. Question 4 asked students to articulate what parts of being a teacher had become more valuable to them. Four distinct themes emerged from students' responses to this question*:* (a) making a difference, (b) uncovering potential in students/lighting fires, (c) patience and respect, and (d) student–teacher rapport. Question 5 asked students to articulate what parts of being a teacher had become less valuable to them. Again, four distinct themes emerged from students' responses to this question*:* (a) the environment/working conditions, (b) teacher disengagement with the profession, (c) motivation, and (d) respect. The last question of the interviews was open-ended, asking students to share any other aspects of their service learning experience which shaped their attitudes toward the profession. Two main themes emerged out of students' responses: motivation and respect.

## Part Two: Emerging Themes From Reflective Case Study Assignment

Respondents chose three main themes for which to respond: fear, identity and integrity, and community. The reflective component of these case studies explored these themes in relation to the participants' tutoring experiences. The requirements and criteria for this assignment can be found in Appendix B. Each student was asked to make connections between their identified theme from Palmer's text (1998) and his or her tutoring experiences at the urban alternative school.

## DISCUSSION

The voices of the participants in this study may provide information about how to reorganize teacher education experiences so we can help students make fully informed decisions to enter the profession. Service learning provides a practical application for learning, opportunities for reflecting on one's activities, and for providing a bridge between theory and practice. Both kinds of data, the student interviews and the reflective component of the case studies, explored the cognitive and affective dimensions of what it means to be a quality teaching professional.

In terms of the demands of the teaching profession, the themes emerging from the student interview data indicate the profession is challenging, there is a lack of respect for teachers, and teachers are undervalued. These themes reflect some of what we already know about the demands of the teaching profession. Though the specific term *burnout* did not emerge in the results of this study, teachers who are unable to provide for their students both physically and emotionally experience emotional exhaustion and eventually burnout (Maslach, Jackson, & Leiter, 1996). Grayson and Alvarez (2007) asserted that emotional exhaustion in teachers is a "tired feeling that develops over time as one's emotional resources are drained" (p. 1350). There is a degree of external validity to the findings in connection with the literature on teacher education. Because there are myriad affective dimensions in the teaching profession, centering on the idea of teaching to the whole person, future teachers will need to draw on their emotional resources.

Themes demonstrating the rewards of teaching emerged from the participants' interview data. These themes include making a difference in students' lives, uncovering potential in students, patience and respect with students, and good student teacher rapport. All of these themes reflect the affective dimensions of the teaching profession because they involve aspects of caring about a student's progression as a learner and development as a person. Personal satisfaction in teaching is particularly important, in part, because most teachers enter the profession so that they can make differences in their students' lives. It is unlikely that they enter the profession for the financial rewards.

Students wrote about the four themes of fear, identity and integrity, and community in the reflective component of their case studies. Respondents made meaning of the affective dimensions of these themes in the context of their work with urban alternative school students. Respondents demonstrated deepened learning of the fears inherent in the teaching and learning space. Not all fears are negative. Teachers need to be aware of what is going on with their students, have empathy for them, and conduct themselves as positive role models. The importance of identity and integrity was navigated and explored; respondents in this study reinforced the value of prior experiences for both teacher and student. Part of this valuing of prior experiences grows through mutual respect between and among teachers and learners. The theme of community, articulated through the respondents' case studies, illustrated how the creation of a trusting environment can promote and deepen learning. All of these themes reflect the affective dimensions important in the profession of teaching.

Future teachers need to understand both the demands and rewards of teaching and recognize the stresses of teaching, including dealing with disengaged students, disconnected teachers, and the political realities of schools (Palmer, 1998). The intention underpinning service learning work is not to accentuate the negative dimensions or discount the rewarding parts of being a teaching professional, but to prepare preservice teachers' entry into the profession with a realistic perception of the work that awaits them should they decide to continue on and work toward licensure. Nothing can replace practical experience in the K–12 urban classroom for making an informed decision to enter the profession of teaching. Teaching is important work, so the conscious decision to become a teacher should be fully investigated and at least partially informed by practical experiences. Similarly, people who are preparing for STEM careers profit from having experiences in STEM fields so they learn within the authentic context of the work that awaits them. Service learning opportunities, for example, in science, provide opportunities for civic engagement and developing ethics in society. Students need experiential education such as service learning opportunities to become active and engaged citizens whose help can generatively and ethically shape our democratic society.

## Emotions, Learning, and Service Learning: Making a Case for STEM Fields

Until recently, approaches to learning and foundations for pedagogical practice in service learning have paid little attention to the rich possibilities and connections between emotions and learning in service learning contexts. Research from cognitive psychology and from cognitive neuroscience have demonstrated again and again that emotion can facilitate learning through myriad mechanisms (Immordino-Yang, 2015). There is an optimal level of emotional engagement, or arousal, for achieving peak performance across a wide variety of contexts (Yerkes & Dodson, 1908), including classroom learning (Naude, van den Bergh, & Kruger, 2014). The relationship between arousal and learning can be described using an inverse U-shaped function, where low levels of arousal lead to disengaged, disinterested students who are unwilling to try hard; moderate levels of arousal lead to attentive, interested students who employ effort; and high levels of arousal lead to distracted, overwhelmed students who develop anxiety and stress. We propose that service learning experiences are an effective way to bring students up to the optimal, moderate level of arousal for learning.

At a basic level, emotional information is found to engage and sustain attention better than nonemotional information (e.g., Weinberg, Ferri, & Hajcak, 2013; Yiend, Barnicot & Koster, 2013). Being able to attend to something well is the first step toward encoding that information into a lasting memory, or learning (Atkinson & Shiffrin, 1968). Emotional information is better remembered than nonemotional information at later recall stages (Bennion, Ford, Murray, & Kensinger, 2013). While not commonly discussed as such, interest or curiosity can be considered emotional experiences (Silvia, 2008). In this framework, engaging a student's curiosity about or interest in course material will bring their emotional engagement levels up to a range that facilitates learning and performance.

The need to help students approach and engage course material in a nonthreatening way is particularly important in STEM fields, where pedagogical methodologies drawing from social-emotional learning have not been as widely adopted (Wood, 2009). Minorities and females have shown higher rates of science anxiety and lower rates of participation in STEM areas (e.g., Brainard, 2007; Udo, Ramsey, & Mallow, 2004), underscoring the importance of using pedagogical techniques that make course material more approachable. Others have argued for the employment of service learning in STEM education to increase student success (Hoachlander & Yanofsky, 2011; Immordino-Yang & Faeth, 2010; Willis, 2010). Service learning has been found to increase student understanding of course content in an engineering course (Tsurusaki & Anderson, 2010). Overall, service learning and other hands-on-based curricular activities have been found to increase students' gains in STEM classes (Thiry, Laursen, & Hunter, 2011). Perhaps connecting STEM content to real-life experience through service learning reduces students' science and mathematics anxiety by using a comprehensible, experience-based arena rather than a less-tangible, textbook-based arena.

As is evidenced by the service learning project discussed here and elsewhere (see Bringle & Hatcher, 1999; Kuh, 2003; Strage, 2000), providing students with opportunities to use their course material in the field increases their interest, facilitating their ability to maintain their attention to, encode, and later recall course material. Research has shown that positive mood states are associated with increases in creativity (for a meta-analysis, see Davis, 2009). It is also likely that service learning can help reduce some of the content-specific anxiety associated with course learning, particularly in STEM classes.

Service learning challenges students to fit their understanding of their fields to applied settings, both practically and morally. Every profession has moral guidelines taught in classroom settings, be it the ethics of conducting research, the ethics of applying science, or the ethics of educating others. Service learning experiences provide an opportunity for students to apply textbook ethics and moral guidelines in real-life scenarios. As Parker Palmer (2014) stated,

> when I look at the malfeasance of well-educated leaders in business and finance, in health care and education, in politics and religion, I see too many people whose expert knowledge—and the power that comes with it—has not been joined to a professional ethic, a sense of communal responsibility, or even simple compassion. (p. vii)

Pedagogical innovations involving reflective practice in service learning are now being explored through empirical studies, ones that acknowledge and privilege the natural interplay of the intellectual and the emotional (Felten, Gilchrist, & Darby, 2006; Immordino-Yang & Damasio 2007; Shapiro, Brown, & Astin, 2011). Reflective practices within service learning contexts provide another avenue for increasing student interest and positive experience with course materials. These practices can help students to understand not only how their area of study fits within the larger context of nonacademic communities, but how they fit as educated ambassadors of their fields within that larger framework.

## APPENDIX A
## Hall's (2009) Service Learning Case Study Questions

### Student Interview Questions

1. Through the experiences in your tutoring and teacher observations, what is your perception of the teaching profession as a whole? Have these perceptions changed? If so, how? If not, how have your experiences confirmed your existing perception(s)?
2. What positive experiences in your prepracticum have contributed to your interest in becoming a teacher?
3. What experiences in your prepracticum have detracted from or lessened your interest in becoming a teacher?
4. What parts of being a teacher have become more valuable to you?
5. What parts of being a teacher have become less valuable to you?
6. Is there anything else that you would like to share about your prepracticum experiences that have affected your attitudes toward teaching? If so, what and how?

## APPENDIX B
## HALL'S (2009) REFLECTIVE COMPONENT OF THE CASE STUDY

Reflections on a theme in Parker Palmer‘s (1998) *The Courage to Teach* in two parts.

*Directions:* Choose a theme of interest to you from Palmer‘s text in either Chapter 1 or 2 and describe how this theme is important to you in your (current and future) teaching life. Make sure you:

1. *Part 1: Clearly articulate a theme from Palmer's text (from either Chapter 1 or 2)*
   - Use at least two quotes that illustrate this theme. These quotes come directly from the text (remember to cite page numbers).
   - Explain why this theme is important to you in your current and future teaching life.
2. *Part 2: Making meaning of your service learning experience*
   - Using the same theme from Palmer's work that you chose above, describe how this theme is important in your work with your student.
   - Give examples of how this theme informs your tutoring experiences at the urban alternative school.

## REFERENCES


Atkinson, R. C., & Shiffrin, R. M. (1968). Human memory: A proposed system and its control processes. In K. W. Spence & J. T. Spence (Eds.), *The psychology of learning and motivation* (pp. 89–195). New York, NY: Academic Press.

Aurobindo. (1950). *The ideal of the Karmayogin.* Sri Aurobindo Press.

Ball, D., & Cohen, D. (1999). Developing practice, developing practitioners: Toward a practice-based theory of professional education. In L. Darling-Hammond & G. G. Sykes (Eds.), *Teaching as the learning profession: Handbook of policy and practice* (pp. 1–32). San Francisc, CA: Jossey-Bass.

Bénabou, R., & Tirole, J. (2003). Intrinsic and extrinsic motivation. *Review of Economic Studies, 70*(3), 489–520.

Bennion, K. A., Ford, J. H., Murray, B. D., & Kensinger, E. A. (2013). Oversimplification in the study of emotional memory. *Journal of the International Neuropsychological Society, 19,* 953–961.

Berman, S. (2006). *Service learning: A guide to planning, implementing, and assessing student projects.* Thousand Oaks, CA: Corwin Press.

Brainard, J. (2007). The tough road to better science teaching. *Chronicle of Higher Education,* 16–18.

Bringle, R. G., & Hatcher, J. A. (1999, Summer). Reflection in service learning: Making meaning of experience. *Educational Horizons, 77*(4), 179–185.

Chickering, A. (2006). Strengthening spirituality and civic engagement in higher education. *Journal of College and Character, 8*(1). Retrieved March 2, 2007, from http://www.tandfonline.com/doi/pdf/10.2202/1940-1639.1153

Cronon, W. (1998) Only connect: The goals of a liberal education. *The American Scholar 67*(4), 73–80.

Darling-Hammond, L. (1998). Teachers and teaching: testing policy hypotheses from a national commission report. *Educational Researcher, 27*(1), 5-15.

Darling-Hammond, L. (2007). The flat earth and education: How America's commitment to equity will determine our future. *Educational Researcher, 36*(6), 318–334.

Davis, B., & Sumara, D. (2002) Constructive discourses and the field of education: Problems and possibilities. *Educational Theory, 52*(4), 409–428.

Davis, M. A. (2009). Understanding the relationship between mood and creativity: A meta-analysis. *Organizational Behavior and Human Decision Processes, 108,* 25–38.

Dewey, J. (1899) *The school and society: Being three lectures by John Dewey, Supplemented by a Statement of the University Elementary School.* Cambridge, England: Cambridge University Press.

Eckel, P. D., & Kezar, A. J. (2003). *Taking the reins: Institutional transformation in higher education.* Westport, CT: Greenwood.

Eyler, J. S. (2001, Summer). Creating your reflection map. *New Directions for Higher Education, 114,* 35–43.

Felten, P., Gilchrist, L. Z., & Darby, A. (2006). Emotion and learning: Feeling our way toward a new theory of reflection in service-learning. *Michigan Journal of Community Service Learning, 12*(2), 38–46.

Grayson, J. L., & Alvarez, H. K. (2008). School climate factors relating to teacher burnout: A mediator model. *Teaching and Teacher Education, 24*(5), 1349–1363.

Hall, M. P. (2005). Bridging the heart and mind: Community as a device for linking cognitive and affective learning. *Journal of Cognitive Affective Learning, 1*, 8–12.

Hall, M. P. (2007). Service learning and intentionality: Creating and assessing cognitive-affective learning connections. *Journal of Higher Education Outreach and Engagement, 11*, 33–43.

Hall, M. P. (2009, June). Service learning in urban alternative schools: Investigating affective development in preservice teacher education. *Transformative Dialogues: Teaching and Learning Journal, 2*(3). Retrieved from https://www.researchgate.net/publication/264860358_Service_Learning_in_Urban_Alternative_Schools_Investigating_Affective_Development_in_Preservice_Teacher_Education

Hall, M. P. (2012, December). Consultation with the mind: The power of reflective writing for deepened learning and self-knowledge in teacher education. *Dev Sanskriti International Interdisciplinary Journal, 1*(1).

Hoachlander, G., & Yanofsky, D. (2011). Making STEM real. *Educational Leadership, 68*(6), 1–6.

Immordino-Yang, M. H. (2015). *Emotions, learning, and the brain: Exploring the educational implications of affective neuroscience.* New York, NY: Norton.

Immordino-Yang, M. H., & Damasio, A. R. (2007). We feel, therefore we learn: The relevance of affective and social neuroscience to education. *Mind, Brain and Education, 1*(1), 3–10.

Immordino-Yang, M. H., & Faeth, M. (2009). The role of emotion and skilled intuition in learning. In D. A. Sousa (Ed.), *Mind, brain, and education* (pp. 66–81). Bloomington, IN: Solution Tree Press.

Immordino-Yang, M. H., & Faeth, M. (2010). The role of emotion and skilled intuition in learning. In D. A. Sousa (Ed.), *Mind, brain, & education: Neuroscience implications for the classroom* (pp. 66–81). Bloomington, IN: Solution Tree Press.

Klein, J. T. (2005). *Humanities, culture, and interdisciplinarity: The changing American academy.* Albany: State University of New York Press.

Kuh, G. D. (2003). What we're learning about student engagement from NSSE. *Change, 35*(2), 24–32.

Magolda, M. B. (Ed.). (2000). *Teaching to promote intellectual and personal maturity: Incorporating students' worldviews and identities into the learning process.* San Francisco, CA: Jossey-Bass.

Maslach, C., Jackson, S. E., & Leiter, M. P. (1996). *Maslach burnout inventory manual.* Mountain View, CA: CPP.

McKay-Jackson, C. (2014). A critical approach to social emotional learning instruction through community-based service learning. *Journal of Transformative Education, 12*(3), 292–312.

Meiklejohn, J., Phillips, C., Freedman, M. L., Griffin, M. L., Biegel, G., Roach, A., ... Isberg, R. (2012). Integrating mindfulness training into K–12 education: fostering the resilience of teachers and students. *Mindfulness, 3*(4), 291–307.

Miles, M. B., & Huberman, A. M. (1994). *Qualitative data analysis: An expanded sourcebook.* Thousand Oaks, CA: Sage.

Naude, L., van den Bergh, T. J., & Kruger, I. S. (2014). Learning to like learning: An appreciative inquiry into emotions in education. *Social Psychology of Education, 17,* 211–228.

Ornstein, A. C., Behar-Horenstein, L. S., & Pajak, E. F. (Eds.). (2003). *Contemporary issues in curriculum* (3rd ed.). Needham Heights, MA: Allyn and Bacon.

Owen-Smith, P. (2004, Fall). What is cognitive-affective learning (CAL)? *Journal of Cognitive Affective Learning, 1*(1), 11. Retrieved from https://cndls.georgetown.edu/teachtoconnect/resources/What_is_CAL.pdf

Palmer, P. J. (1998). *The courage to teach: Exploring the inner landscape of a teacher's life.* San Francisco, CA: Jossey-Bass.

Palmer, P. J. (2014). Foreword. In D. P. Barbezat & M. Bush (Eds.), *Contemplative practices in higher education: Powerful methods to transform teaching and learning.* Hoboken, NJ: Wiley.

Piaget, J. (1954). *Construction of reality in the child.* London, England: Routledge/Kegan Paul.

Shapiro, S., Brown, K., & Astin, J. (2011). Toward the integration of meditation into higher education: A review of research evidence. *Teachers College Record, 113*(3), 493–528.

Shepard, L. A. (2000). The role of assessment in a learning culture. *Educational Researcher, 29*(7), 4–14.

Silvia, P. J. (2008). Interest: The curious emotion. *Current Directions in Psychological Science,17*(1), 57–60.

Solley, B. A. (2007). On standardized testing: An ACEI position paper. *Childhood Education, 84*(1), 31–37.

Strage, A. A. (2000). Service learning: Enhancing student learning outcomes in a college-level lecture course. *Michigan Journal of Community Service Learning,* 5–13.

Thiry, H., Laursen, S. L., & Hunter, A. B. (2011). What experiences help students become scientists? A comparative study of research and other sources of personal and professional gains for STEM undergraduates. *The Journal of Higher Education, 82*(4), 357–388.

Tsurusaki, B. K., & Anderson, C. W. (2010). Students' understanding of connections between human engineered and natural environmental systems. *International Journal of Environmental & Science Education, 5*(4), 407–433.

Udo, M. K., Ramsey, G. P., & Mallow, J. V. (2004). Science anxiety and gender in students taking general education science courses. *Journal of Science Education and Technology,13*(4), 435–446. doi:1059-0145/04/1200-0435/0

Vygotsky, L. S. (1978). *Mind in society: The development of higher psychological processes.* Cambridge, MA: Harvard University Press.

Weinberg, A., Ferri, J., & Hajcak, G. (2013). Interactions between attention and emotion: Insights from the late positive potential. In M. D. Robinson, E. Watkins & E. Harmon-Jones (Eds.), *Handbook of cognition and emotion* (pp. 35–54). NewYork, NY: Guilford Press.

Willis, J. (2010). The current impact of neuroscience on teaching and learning. In D. A. Sousa (Ed.), *Mind, brain, & education: Neuroscience implications for the classroom* (pp. 45–65). Bloomington, IN: Solution Tree Press.

Wood, W. B. (2009). Innovations in teaching undergraduate biology and why we need them. *Annual Review of Cellular Developmental Biology, 25,* 93–112.

Yerkes, R. M., & Dodson, J. D. (1908). The relation of strength of stimulus to rapidity of habit-formation. *Journal of Comparative Neurology and Psychology, 18*, 459–482.

Yiend, J., Barnicot, K., & Koster, E. H. W. (2013). Attention and emotion. In M. D. Robinson, R. E. Watkins, & R. Harmon-Jones (Eds.), *Handbook of cognition and emotion* (pp. 97–116). New York, NY: Guilford Press.

Zajonc, A. (2006). Love and knowledge: Recovering the heart of learning through contemplation. *Teachers College Record, 108*(9), 1742–1759. Retrieved from http://www.contemplativemind.org/admin/wp-content/uploads/2012/09/zajonc-love-and-knowledge.pdf

# ABOUT THE CONTRIBUTORS

**Sally Blomstrom** is an associate professor in the Humanities and Communication Department at Embry-Riddle Aeronautical University in Daytona Beach, Florida, where she teaches speech and technical report writing. She has been active in STEM service learning outreach projects for 10 years. Her research focuses on STEM literacy, service learning, and assessment. She has published in the *Journal on Excellence in College Teaching, The International Journal of Research on Service-Learning and Community Engagement, Best Practices in Experiential and Service Learning in Communication,* and *Scholarship for Sustaining Service-Learning and Civic Engagement.*

**Micah N. Bruce-Davis** is an assistant professor at the University of Louisiana at Lafayette. She received her doctorate in Educational Psychology with a focus on gifted education and talent development from the University of Connecticut in 2013; Master of Arts in Curriculum and Instruction from Louisiana State University in 2007; and her Bachelor of Arts in Elementary Education from the University of Louisiana at Lafayette in 2002. She taught K–12 for eight years in southeastern Louisiana. Her research interests include specialized schools, motivation, and the affective needs of gifted students.

**Heidi B. Carlone** is a professor of science education at the University of North Carolina at Greensboro. Carlone's research uses theoretical lenses from anthropology of education, sociocultural theories, and cultural studies to understand short- and long-term impacts of innovative elementary science, environmental, and engineering instruction on diverse K–12 students' identity work.

*Science and Service Learning,* pages 241–247

**Angela Chapman** is an assistant professor of science education in the Department of Teaching and Learning in the College of Education and P–16 Integration at the University of Texas Rio Grande Valley. Her research and scholarship focus on equitable and effective practices in science education, specifically how all students can have access to quality science learning experiences. One aspect of her research focuses on developing curricula that helps students develop a deeper understanding of science vocabulary using personal, cultural, and linguistic capital. Another aspect of her research is investigating factors that influence student perceptions of scientists and how they develop their science identity. Angela has served as a strand coordinator for NARST and has published empirical studies related to her research.

**Kate Corwin** is an educator, business leader, and advocate for disadvantaged urban youth. In 2007, after a successful career as an IT engineer in telecommunications and 9 years as an entrepreneur, Kate founded Green Works in Kansas City, Missouri. Kate envisioned Green Works as both an educational as well as an economic development tool to connect urban youth to the environmental movement and to growing green jobs. Kate is an active community member, having served as chair of the 43rd and Main TIF Advisory Board, on the Landmarks Commission, the Zoning Steering Committee and as president of the Southmoreland Neighborhood Association. She currently serves on the board of MainCor and writes a blog on community development. Kate was named an American Graduate Champion by KCPT, is a recipient of an Environmental Excellence Award from Bridging the Gap, and was named Eco Activist of the Year by *The Pitch* magazine. Kate has worked on campaigns to limit bottled water at City Hall, require recycling at events in city parks, and discourage the planting of ornamental pear trees. Kate is a member of Kansas City Tomorrow and International Women's Forum. She holds undergraduate degrees in Information Technology and Finance and an MA in Economics.

**Sharon Ellerton** is an associate professor in the Department of Biological Sciences at the City University of New York (CUNY), Queensborough Community College (QCC) and has served as faculty liaison/high-impact practices coordinator of academic service learning (ASL) at QCC since 2009. In that role, Dr. Ellerton leads ASL faculty development workshops and researches the impact of ASL on community college students. Results of this research have been published in Palgrave Macmillan's book, *Service Learning at the American Community College, Theoretical and Empirical Perspectives,* and in several journal articles. Dr. Ellerton is a reviewer for the Undergraduate *Journal of Service-Learning & Community Based Research* and mentors students submitting articles. In addition, Dr. Ellerton is a practitioner of ASL, incorporating this pedagogy into her own coursework.

**Allan Feldman** is a professor of science education in the Teaching and Learning Department of the College of Education at the University of South Florida. His scholarship focuses on science teacher education and in particular how practicing science teachers learn from their practice in a variety of subjects including physics, environmental education, and education for sustainability in formal and informal settings. He is co-editor of the book *Educating Science Teachers for Sustainability,* which was published in 2015 by Springer. He has written a number of papers on action research and self-study as well as on science teacher education. Much of his scholarship has been on teacher learning through action research and self-study and through participation in scientific research groups. He is currently co-editor of *Educational Action Research* and has published theoretical and empirical studies of action research and self-study.

**Anita Cuttita Ferdenzi** is a professor in the Department of Social Sciences at the City University of New York (CUNY), Queensborough Community College. Dr. Ferdenzi has served as the principal investigator for the NASAIMUSPIN grant to promote a greater understanding of the mathematics associated with the study of astronomy among pre-service teachers who are representative of minority populations. Her research focus has been the integration of innovative teaching and learning strategies in math and science designed to increase achievement and efficacy for minority groups historically underrepresented among the ranks of science and math teachers. Dr. Ferdenzi has also led an interdisciplinary team of faculty to investigate the effects of constructivist pedagogy on preservice community college students' self-beliefs about math, physics, and metacognition of self-regulated learning in a cornerstone Physics/Education Learning Community. She has also collaborated with mathematics faculty to design and deliver an interdisciplinary cognitive foundations of secondary mathematics course for the NSF grant entitled the TIME QCC—An initiative to increase the number of teachers of secondary mathematics entering the workforce.

**Jen Foreman** currently teaches in the Psychology Department and Education Studies Program at Allegheny College. She graduated from the University of Connecticut's PhD program in educational psychology in 2013. She worked as a research associate for the National Research Center on the Gifted and Talented, a consultant, and a program evaluator prior to that. Jen's research interests include service learning and character development, academically advanced students, neurodiversity, and alternate forms of schooling.

**Cindy M. Gilson** is an assistant professor at the University of North Carolina at Charlotte. She earned her PhD in educational psychology from the University of Connecticut in 2014. Her research interests include gifted and talented education, differentiated curriculum and instruction, classroom discourse, and professional development.

**E. Jean Gubbins** is a professor of educational psychology and an associate director of the National Research Center on the Gifted and Talented at the University of Connecticut. Her research focuses on the differentiation and enrichment of curricula, program development, and professional development.

**Maureen P. Hall** earned her PhD in 2003 at the University of Virginia and is a professor of education in the STEM Education and Teacher Development Department at UMass Dartmouth, where she has been a teacher educator for more than 13 years. Before that, she taught English and social studies for 10 years in secondary classrooms in New Hampshire. Her work bridges literacy, mindfulness, and service learning. Over 2010–2011, Dr. Hall spent 7 months in India as a Fulbright-Nehru Senior Research Scholar, and she currently maintains a visiting professorship at Dev Sanskriti University in Haridwar, Uttarakhand, India. Her first book, *Transforming Literacy* (2011), was written with lead author, Dr. Robert P. Waxler. She also has a forthcoming book in 2017 entitled, *Writing From the Inside (Equinox),* with Dr. Olivia Archibald.

**Barbara Hayford** attended Colorado State University, Pueblo as an undergraduate and graduate student, earning an MS degree in Applied Natural Science. She earned a PhD in entomology with emphasis on aquatic insects at University of Kansas. Dr. Hayford completed a 2-year postdoctoral fellowship/Assistant Directorship of the Central Plains Center for Bioassessment in Lawrence, Kansas, and worked for a year as a project scientist under Mike Johnson at the UC Davis John Muir Institute for the Environment. She currently is a full professor at Wayne State College in Nebraska, where she teaches biology courses and served as director of the A. Jewell Schock Museum of Natural History from 2008 to 2015. Her pedagogical research focuses on ways to effectively evaluate literacy in science, technology, engineering, and math courses, particularly through service learning projects.

**Lacey Huffling** is an assistant professor in the Teaching and Learning Department at Georgia Southern University. Dr. Huffling primarily teaches in the Masters of Arts in Teaching for Middle Grades and Secondary certification. As a science educator, she advocates for equity and access for all learners. In particular, she works with all ages to instill a love of nature and a care for the environment.

**Christie Klimas** is an assistant professor in DePaul's Environmental Science and Studies department where she teaches ecological economics, sustainability, life cycle assessment, study abroad courses on Fair Trade, and ecology. Her research brings together ecology and economics, in urban and tropical settings, to address questions of sustainable resource use.

**Junfei Lu** is an assistant professor in the Rehabilitation Counseling Program at the University of Alabama. He earned his doctorate from the Program of Rehabilitation and Counselor Education at the University of Iowa. His research focuses include wellness and resilience.

**Catherine E. Matthews** is a professor emerita of K–12 science education and environmental education in the Department of Teacher Education and Higher Education at the University of North Carolina Greensboro. She has served as the principal investigator on the NSF-funded Herp Project (Herpetology Education in Rural Places and Spaces) for the past 5 years. Her passion is natural history and her best teaching is done in the out-of-doors.

**James Montgomery** is an associate professor and former chair of the Department of Environmental Science and Studies at DePaul University in Chicago. He is currently interim chair of the Department of Health Sciences at DePaul. He holds a BS and an MS degrees in geology from Baylor University, and a PhD in soil science from Washington State University. He teaches courses in environmental soil science, introductory environmental science courses for nonscience majors, and courses in research methods and environmental impact analysis for environmental science and studies majors. He has also taught a geology course for in-service middle-grade teachers in the Chicago Public Schools (CPS) system as part of DePaul's MS in science education degree program, as well as a forensic science course for CPS middle-grade students. His academic interests include wetland science, urban soils, campus sustainability, and science education.

**Lori Mumpower** is an associate director for the Center for Teaching and Learning Excellence at Embry-Riddle Aeronautical University. Her work with faculty includes developing hybrid courses, directing the campus's digital studio, and providing embedded support to faculty in the College of Arts and Sciences. She has taught a variety of writing and rhetoric courses over the past 15 years and directed a first-year writing program. Dr. Mumpower has always had an interest in service learning in writing courses and currently works with a learning community of ERAU faculty to promote and support service learning on campus.

**Jane L. Newman** is an associate professor and coordinator of the Gifted and Talented Program, Department of Special Education in the College of Education at University of Alabama. She holds a doctorate in educational leadership and administration/special education/gifted and talented. She has spent the majority of her career in public schools (teacher of gifted, director of instruction, high school principal, and assistant superintendent). Her interests include creative productivity, underserved populations and resilience, service learning/community engagement, evidence-based effects of professional development on student achievement. In 2015, she was honored by receiving the UA Award for Outstanding Faculty Initiated Community Engagement Effort. In 2014 she was awarded the UA College of Education's Outstanding Faculty Teaching Award.

**Aminda O'Hare** is an assistant professor of psychology at UMass Dartmouth, where she runs the CAPES Lab (Cognitive and Affective Psychophysiology and Experimental Science Lab). Dr. O'Hare broadly studies the intersection of emotion and cognition using tools of neuroscience. Specifically, she focuses on the factors that lead to emotion facilitating attention and decision making versus interfering with these processes. Recently, she has been investigating the moderating effect of mindfulness meditation on emotion-cognition interactions.

**Alexis Petri** is director of interdisciplinary training at the University of Missouri–Kansas City Institute for Human Development, a University Center of Excellence on Disabilities. With over 15 years of experience working in higher education and 5 years of experience working on access issues related to higher education, Dr. Petri has extensive experience with transition to college and careers for nontraditional students as well as taking leadership roles in building and directing universitywide high-impact learning programs and the campus-community partnerships that are their foundation, such as service learning and internships. Currently, Dr. Petri is co-principal investigator and project director for two projects funded by the National Science Foundation. Dr. Petri also serves as director of service learning for the University of Missouri of Kansas City, campuswide. As part of her research, she conducts program evaluations, writes curriculum, and builds new media/cloud-based learning environments for various campus-community partnerships related to service learning, internships, urban education, community engagement, people with disabilities, and veterans.

**Howard Rosing**, PhD, is a cultural anthropologist and the executive director of the Steans Center for Community-Based Service Learning and Community Service Studies at DePaul University. He is also co-director of DePaul's graduate program in sustainable urban development (SUD). His research focuses on urban food access, economic restructuring, community food systems, and food justice movements in Chicago and the Dominican Republic. He is co-developer of the Chicago Urban Agriculture Mapping Project (cuamp.org), a public resource for documenting Chicago's food production.

**Vazgen Shekoyan** is an assistant professor in the Department of Physics at the City University of New York (CUNY), Queensborough Community College. Dr. Shekoyan did his PhD in physics education research at Rutgers University. His PhD thesis was related to introduction of ill-structured problems into physics classes to enhance both students' understanding of the content material and their cognitive abilities (especially, students' epistemic cognition). As a co-principal investigator for the NSF TUES grant, Dr. Shekoyan evaluated the impact of an active-learning gemology studio course on students' learning, science skills, interest in STEM, as well as various affective factors such as motivation and self-efficacy. He is currently involved in a number of pedagogical projects exploring the impact of various STEM classroom innovations and teaching approaches on student learning.

**Terry Tomasek** is an associate professor in the Teacher Education Department at Elon University. As a science educator, she works with all ages of people to involve them in multiple levels of science engagement—from taking 5 year olds on nature hikes to discover and appreciate worms, to working with high school students on extensive aquatic turtle population studies, to working with adults interested in participation in citizen science. Dr. Tomasek also teaches science pedagogy, interdisciplinary science, and global studies courses at Elon University in central North Carolina.

Printed and bound by CPI Group (UK) Ltd, Croydon, CR0 4YY
15/09/2026
14962338-0003